Into the woods

MANCHESTER
1824
Manchester University Press

New
Ethnographies

Series editor
Alexander Thomas T. Smith

Already published

An ethnography of NGO practice in India: Utopias of development Stewart Allen

The British in rural France: Lifestyle migration and the ongoing quest for a better way of life Michaela Benson

Ageing selves and everyday life in the North of England: Years in the making Catherine Degnen

Salvage ethnography in the financial sector: The path to economic crisis in Scotland Jonathan Hearn

Occupational health and social estrangement in China Wing-Chung Ho

Chagos islanders in Mauritius and the UK: Forced displacement and onward migration Laura Jeffery

South Korean civil movement organisations: Hope, crisis and pragmatism in democratic transition Amy Levine

Integration in Ireland: The everyday lives of African migrants Fiona Murphy and Mark Maguire

Environment, labour and capitalism at sea: 'Working the ground' in Scotland Penny McCall Howard

An ethnography of English football fans: Cans, cops and carnivals Geoff Pearson

Iraqi women in Denmark: Ritual performance and belonging in everyday life Marianne Holm Pedersen

Loud and proud: Passion and politics in the English Defence League Hilary Pilkington

Literature and agency in English fiction reading: A study of the Henry Williamson Society Adam Reed

International seafarers and transnationalism in the twenty-first century Helen Sampson

Tragic encounters and ordinary ethics: The Palestine-Israel Conflict in British universities Ruth Sheldon

Devolution and the Scottish Conservatives: Banal activism, electioneering and the politics of irrelevance Alexander Smith

Exoticisation undressed: Ethnographic nostalgia and authenticity in Emberá clothes Dimitrios Theodossopoulos

Immersion: Marathon swimming, embodiment and identity Karen Throsby

Enduring violence: Everyday life and conflict in eastern Sri Lanka Rebecca Walker

Performing Englishness: Identity and politics in a contemporary folk resurgence Trish Winter and Simon Keegan-Phipps

Into the woods

An epistemography of climate change

Meritxell Ramírez-i-Ollé

Manchester University Press

Published by Manchester University Press
Altrincham Street, Manchester M1 7JA
www.manchesteruniversitypress.co.uk

British Library Cataloguing-in-Publication Data
A catalogue record for this book is available from the British Library

ISBN 978 1 5261 4098 2 hardback

First published 2020

Typeset
by Toppan Best-set Premedia Limited
Printed in Great Britain
by TJ International Ltd, Padstow

I do not believe that social science will 'save the world' although I see nothing at all wrong with 'trying to save the world' – a phrase which I take here to mean the avoidance of war and the re-arrangement of human affairs in accordance with the ideals of human freedom and reason.
Such knowledge as I have leads me to embrace rather pessimistic estimates of the chances.
But even if that is where we now stand, still we must ask: If there *are* any ways out of the crises of our period by means of intellect, is it not up to the social scientist to state them?

Charles Wright Mills, The Sociological Imagination *([1959] 2000: 193).*

Many theories of knowledge are morality plays set in a Manichaean cosmos. The source of light is experience; its agent 'reason'. The source of darkness is culture; its agent authority. The remaining *dramatis personae* are garbed according to their origins. Truth, validity, rationality, objectivity are to be seen among the many white-apparelled children of the light; error and irrationality, custom, convention, dogma and many others are dressed in black.
The moving principle of the drama is the unremitting conflict of the two opposed and irreconcilable forces. There is nothing to be said in favour of this Manichaean mythology.

Barry Barnes, T. S. Kuhn and Social Science *(1982: 22).*

Contents

Figures

Boxes

Preface: saving climate science

Climate science has long been in trouble and I wish to help it with this book. As the climate scientist Michael E. Mann (2012) vividly recounts in his autobiography, the 'climate wars' and heated public disputes about the accuracy of climate science originated in the early 1990s when the United Nations Intergovernmental Panel on Climate Change (IPCC) published its first report summarising the scientific evidence of climate change for policy-makers. As Mann also narrates in first person, the most malicious personal attack on climate science occurred in November 2009, when thousands of private emails and documents sent and received by prominent climate scientists (including Mann himself) were stolen and published online. The anonymous hackers justified this ominous attack by saying, 'We feel the climate science is, in the current situation, too important to be kept under wraps. We hereby release a random selection of correspondence, codes and documents. Hopefully, it will give some insights into the science and the people behind it' (Pearce, 2010: 166). The hackers indeed succeeded in opening the workings of climate science to the public. For months, the climate scientists whose emails had been stolen were the focus of media attention and were investigated by multiple university and parliamentary inquiries under allegations of obstruction to open access to scientific data and failures of objectivity in peer-review and research assessment. The hacking and its aftermath, as the House of Commons admitted in its inquiry report, were a 'traumatic and challenging experience for all involved and to the wider world of science' (House of Commons Science and Technology Committee, 2010: 33).

The authority of climate scientists has been eroded since the turn of the twenty-first century by what seems to be a more general phenomenon: what happens inside many sciences has become visible to a highly educated and self-confident citizenry, as television and the Internet have opened up once exclusive and hidden spaces to public scrutiny (Collins, 2014; Gregory and Miller, 1998). The challenge faced by climate scientists is depicted in a cartoon published in *The Economist* shortly after the hacking (Figure P.1): the robust stock of knowledge that has been privately generated and validated by thousands of climate scientists for years (represented by a fortified tower of IPCC reports in the cartoon) is now under direct assault and surveillance from outside experts (as seen by the fact that these outsiders wear laboratory coats in the cartoon).

P.1 Climate scientists and their work have been under intense public scrutiny.

I am of the opinion that there is little that climate scientists, individually or collectively, can do to reverse a broader secular trend affecting the credibility of technical and scientific experts and traditional authorities (Barnes, 2005). The growing 'culture of suspicion' towards climate scientists has to do with: i) broader social changes caused by the expansion of formal education and the increasing accessibility of information, which have given rise to the so-called 'climate sceptical blogosphere' (Sharman, 2014); and ii) the actual changes that have occurred within climate science throughout the twentieth century, the professionalisation of climate science and its increased associations with political institutions, which have meant that climate scientists are not perceived as independent experts (Lahsen, 2013a; Agar, 2012: 397; Edwards, 2010). Consequently, these 'uninvited guests', who continuously show up on the doorsteps of the fortified house of climate science and who cause some inconvenience to its inhabitants, are not likely to disappear in the future. Strategically, I suggest, climate scientists should acclimatise to this new context by making themselves and their work accountable to their sceptical audiences and demonstrating why they are virtuous and competent and why climate science is worthy of public trust and money (O'Neill, 2013, 2002; Jasanoff, 2010; Hulme and Ravetz, 2009; Shapin, 1994).

I have written this book because I worry that climate scientists might not be well equipped to survive future public examinations of their work, not because climate science is not robust enough but because the source of its robustness – the fact that climate science is made by humans – is publicly condemned and dismissed by scientists themselves. I came to this conclusion after analysing the way individual scientists and scientific institutions publicly responded to the allegations made against them after the hacking episode in November 2009 (Ramírez-i-Ollé, 2015a). To my surprise, I discovered that scientists agreed with their critics that the stolen emails were embarrassing. Rather than providing more context for the electronic correspondence – by explaining that scientific facts are a product of human labour and negotiation and that disciplinary commitments, politics and personal relationships have a bearing on scientists' handling, interpretation and reporting of data – some scientists criticised the very social processes and influences that constitute the practical reliability of all sciences. As one physics professor put it, 'Science often falls short of its ideals, and the climate debate has exposed some shortcomings. Science is done by people, who need grants, who have professional rivalries, limited time, and passionately held beliefs. All these things can *prevent us* from finding out what works' (Butterworth, 2010: emphasis added). By upholding a conventional and very false image of the procedures of science, scientists might have inadvertently given weaponry to the critics of climate science who – because of bad faith or genuine ignorance – uphold scientific standards that no science will ever reach. If climate scientists continue romanticising their work (or allow others to do so), they will likely generate further public distrust and cynicism. After all, we should not be surprised that educated and well-informed people look for alternative explanations and experts when things do not turn out to be quite as they were always told.

The story of how things have got to a point at which scientists have surrounded themselves by walls of hype, myth and denial is too long to be told here (see Sarewitz, 2016 and Shapin, 2001 for explanations); I instead aim to bring these walls down and make the now fairly open house of climate science more comprehensible to outsiders. Climate science needs neither heroes nor Public Relations agents to regain its credibility; rather, it needs sociologists, historians, anthropologists and philosophers of science (in short, Science and Technology Studies scholars) who can challenge damaging mythologies about climate science with what I call 'epistemographies of climate change', or empirically rich and contextualised accounts of climate knowledge in the making. This book draws on a long tradition of epistemographic studies in order to tell the story of how, with what confidence and on what grounds, a small group of climate scientists – 'dendroclimatologists' specifically – were able to generate knowledge of climate change in Scotland from the study of the Caledonian forests and to link their specific data to broader trends of global climate change. Ultimately, I hope that, by offering a detailed account of the social life of climate science, readers will grant authority to climate science not because it justifies itself as a self-sufficient worldview or substitute of God, but because, as shown in this book, it is a fine human achievement and our most reliable source of available expertise.

Acknowledgements

Without the financial support from Obra Social Sa Nostra Caixa de Balears, the Economic and Social Research Council and The Sociological Review Limited Foundation, the research that led to this book would never have been possible. The commissioning editor Thomas Dark and the production team at Manchester University Press (Jen Mellor, Robert Byron, Diane Wardle and Humairaa Dudhwala) have done a great job in producing a beautiful book.

Dr Rob Wilson, Dr Miloš Rydval and many other dendrochronologists have tolerated my presence and my questions all these years with admirable openness, kindness and patience. I treasure our friendship and I am thankful for all the wonderful experiences I have shared with them.

Many talented teachers and generous colleagues have accompanied me throughout the ten years it has taken me to complete this book, from the choice of topic to the point of publication. At the Autonomous University of Barcelona, I trained as an undergraduate student in the 'invisible school' of sociology that Professor Joan Estruch i Gibert and Professor Salvador Cardús i Ros established; being part of this collective has had many unintended consequences such as meeting my wonderful husband, Albert Costa, among its disciples. At the University of Edinburgh, I was supervised as a doctoral student by Professor Steve Sturdy and Dr Emma Frow and by the broader subject group of Science, Technology and Innovation Studies; I miss my dearest friends Dr Sara Beà and Dr Thoko Kamwendo. At the Department of Science and Technology Studies at University College London, I found what I was looking for as a recently graduated Doctor: a supportive environment where I could learn from enthusiastic and experienced teachers and colleagues, most notably from Professor Joe Cain and Professor Brian Balmer. The Sociological Review and Keele University offered me what I needed most as an early-career postdoctoral researcher: a salary, freedom, time and expert advice; the latter of which was generously given by Dr Michaela Benson and Professor Joanna Latimer. At FIAC Idiomes, Joe Millanes and Santi Miralda have been kind enough to accommodate my hobbies.

Researching and writing this book has taken away time from being with family and friends. Ivan Kralj suffered the most from my absences; I hope he will see this book as a slight compensation.

Thank you all very much. Moltes gràcies a tothom. Najlepša hvala vsem.

Series editor's foreword

When the *New Ethnographies* series was launched in 2011, its aim was to publish the best new ethnographic monographs that promoted interdisciplinary debate and methodological innovation in the qualitative social sciences. Manchester University Press was the logical home for such a series, given the historical role it played in securing the ethnographic legacy of the famous 'Manchester School' of anthropological and interdisciplinary ethnographic research, pioneered by Max Gluckman in the years following the Second World War.

New Ethnographies has now established an enviable critical and commercial reputation. We have published titles on a wide variety of ethnographic subjects, including English football fans, Scottish Conservatives, Chagos islanders, international seafarers, African migrants in Ireland, post-civil war Sri Lanka, Iraqi women in Denmark and the British in rural France, among others. Our list of forthcoming titles, which continues to grow, reflects some of the best scholarship based on fresh ethnographic research carried out all around the world. Our authors are both established and emerging scholars, including some of the most exciting and innovative up-and-coming ethnographers of the next generation. *New Ethnographies* continues to provide a platform for social scientists and others engaging with ethnographic methods in new and imaginative ways. We also publish the work of those grappling with the 'new' ethnographic objects to which globalisation, geopolitical instability, transnational migration and the growth of neoliberal markets have given rise in the twenty-first century. We will continue to promote interdisciplinary debate about ethnographic methods as the series grows. Most importantly, we will continue to champion ethnography as a valuable tool for apprehending a world in flux.

Alexander Thomas T. Smith
Department of Sociology, University of Warwick

Introduction: epistemography

This book presents an epistemography. While '–graphy' means 'description' or 'study', 'episteme' refers to 'knowledge' or 'skill'. Epistemography, therefore, can be defined as the empirical study of expertise. More specifically, this book is an epistemography of climate change, which means that it is a detailed description and analysis of the production of scientific knowledge of climate change. In the following chapters, I outline the process by which a group of scientists created a graph showing the evolution of temperatures in Scotland over the last 800 years from the analysis of tree growth (Figure 0.1). In their own words (Rydval et al., 2017: 2970), this graph shows that 'Within the context of reconstruction uncertainty, recent summertime warming is not significantly more pronounced than past reconstructed warm periods (e.g. around 1300 and 1500)'. In other words, these scientists concluded that while the Scottish climate has changed substantially over the last 800 years, it was impossible to know for certain from the available data whether the recent warming in Scotland is exceptional. This chapter reviews the tradition of epistemographic studies spanning the fields of history, sociology, anthropology and philosophy of science that have inspired the present epistemography. To a certain extent, the chapter is a reinterpretation of the eclectic and relatively new field of Science and Technology Studies (STS). While being aware of the dangers of essentialising STS, my purpose is to synthesise the existing modes of inquiry in the field (Felt et al., 2017; Sismondo, 2009) into a set of methodological precepts that could be used by scholars from within and beyond STS to study contemporary and historical forms of expertise.

Situated impartiality

The historian of science Peter Dear (2001) coined the neologism 'epistemography' to refer to the descriptive and non-normative approach shared by STS scholars who study what science and technology, as human activities, *actually* are and have been, rather than what they *should* be. Significantly, Dear defined epistemo-*graphy* in opposition to epistemo-*logy*: while the former seeks to make empirical statements about knowledge (what counts as credible knowledge in specific circumstances), the latter aspires to make normative statements (what ought to count as valid

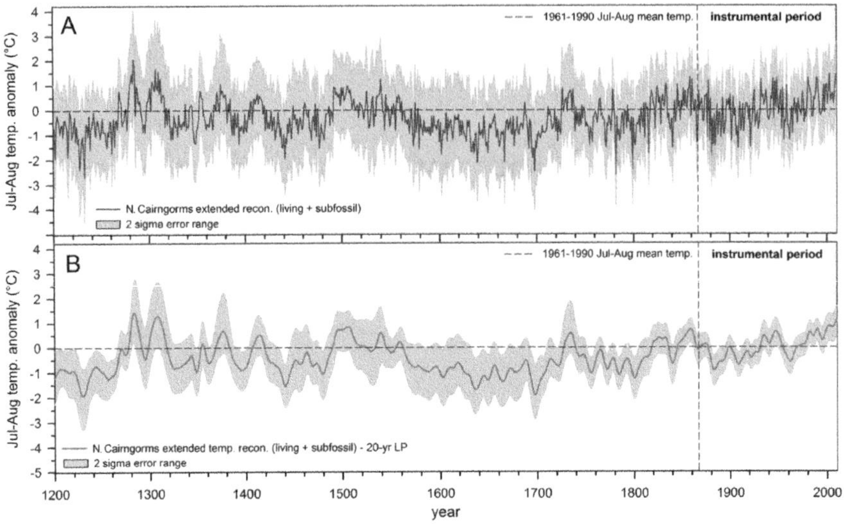

0.1 This epistemography examines the process whereby scientific knowledge of climatic change in Scotland (the graph) was collectively created and approved.

knowledge). In other words, while epistemographers study the bases of *credibility*, epistemologists are concerned about the grounds of *validity*. Dear (2001: 130–131) explained it in this way:

> Epistemography is the endeavor that attempts to investigate science 'in the field', as it were, asking such questions as these: What counts as scientific knowledge? How is that knowledge made and certified? In what ways is it used or valued? 'Epistemography' as a term signals that descriptive focus, much like 'biography' or 'geography'. It designates an enterprise centrally concerned with developing an empirical understanding of scientific knowledge, in contrast to *epistemology*, which is a prescriptive study of how knowledge can or should be made.

Epistemographers are professionally expected to develop a certain discipline and to avoid taking sides in disputes about what counts as 'knowledge'. This is why the reader will not find in this book any discussion from me trying to establish whether Figure 0.1 is true or false. These sorts of claims and debates are the domain of expertise of climate scientists, and are precisely the focus of study of this epistemography. As an epistemographer, my only concern is to offer socio-historical explanations of why Figure 0.1 became accepted by some people at a certain point in time. I have no credited expertise and no interest in contributing to the expert discussions of climate scientists, and I simply accept the scientific consensus on climate science (including Figure 0.1) and seek to explain its formation. The epistemographer's personal opinion on the merits of the ideas under study is not only irrelevant for the epistemographic analysis but will likely be unoriginal as it will reflect 'common-sense' opinions already existing in her or his society. What matters for understanding what a given society comes to 'know' as reality – the purpose of any epistemography – is to analyse the context, and the processes by which ideas emerge and come to be shared collectively in certain circumstances.

The epistemographic approach to the study of whatever passes for 'knowledge' in a society (including the epistemographer's own society) is inspired by the perspective traditionally espoused by sociologists of knowledge. In *The Social Construction of Reality*, Peter Berger and Thomas Luckmann (1966: 15) famously stated, 'The sociology of knowledge must concern itself with whatever passes for "knowledge" in a society, regardless of the ultimate validity and invalidity (by whatever criteria) of such knowledge.' As Barnes has noted (2016: 116), Berger and Luckmann's famous words are strikingly similar to David Bloor's ([1976] 1991: 7) later formulation of the postulates of 'impartiality' and 'symmetry' in the sociological study of scientific knowledge: 'It [SSK] would be impartial with respect to truth and falsity, rationality or irrationality, success or failure. Both sides of these dichotomies will require explanation. It would be symmetrical in its style of explanation. The same types of cause would explain, say, true and false beliefs.'

Importantly, the impartiality principle does not imply any *dis*belief in the knowledge under study. In other words, my epistemography does not deny that climate science offers reliable knowledge and evidence. Doing so would confuse 'methodological relativism', espoused here, with 'epistemic relativism'. As Barry Barnes and David Bloor (1982: 23) clarify, 'Our equivalence postulate [methodological relativism] is that all beliefs are on a par with one another with respect to the causes of their credibility. It is not that all beliefs are equally true or equally false, but that regardless of truth and falsity the fact of their credibility is to be seen as equally problematic.' Precisely, my epistemography acknowledges the greater credibility that climate science currently enjoys over other sources of knowledge of climate change such as, for instance, quotidian experiences of daily weather by analysing the former rather than the latter.

To show that climate science is socially constructed does not imply to conclude that climate change does not exist (Latour, 2004); in fact, social constructivism, as a theory of knowledge, leads to quite the opposite conclusion: climate change is perceived as a real and dangerous phenomenon *partly* thanks to the work done by scientists. As the sociologist Ulrich Beck (1992: 27) explains, environmental hazards 'require the "sensory organs" of science – theories, experiments, measuring instruments – in order to become visible or interpretable as hazards at all'. Consequently, the impartiality principle allows the epistemographer to accept given facts (in my case, acknowledging that scientific accounts of climate change are true and that climate change is a genuine threat) while also offering sociological explanations for their acceptance.

The impartiality principle is neither a suggestion to adopt a 'view from nowhere' nor for being personally and politically detached from the subjects or the object of study. Conceiving the positionality of epistemographers in terms of the dichotomy 'objectivity' vs 'engagement' is, using Casper Bruun Jensen and Peter Lauritsen's (2005: 60) words, a 'badly posed problem'. All forms of objective knowledge – including epistemographies – result from 'people with bodies, situated in time, space, culture, and society, and struggling for credibility and authority' (Shapin, 2010). The feminist philosopher Donna Haraway (1988: 582) coined the concept of 'situated knowledges' to argue that 'objectivity turns out to be about particular and specific embodiment, and definitely not about the false vision promising transcendence of all limits and responsibility'. Haraway (1988: 583, 590) presents a seeming paradox: 'Only partial perspective promises objective vision' and 'the only way to find a larger vision is to be somewhere in particular'. Paraphrasing Haraway, I would argue that the only way to study knowledge-claims impartially and to bypass validity debates is to be a partial insider; only situatedness and partaking promise more accurate accounts of people's beliefs, and, hence, more acceptable and objective epistemographies (see Box 10, p. 108). As I illustrate throughout the book, I was allowed by my subjects to conduct and publish the research for this book precisely because I became intensely involved in their work and lives.

Remaining impartial with regards to the validity of certain knowledge-claims does not prevent epistemographers from contributing to credibility debates if they wish to do so. In my case, as stated in the preface, I have written this book to reassure climate scientists that their practices are justified and to help them to respond to, what I think are, unfair accusations. Crucially, I decided to write this book and adopt a denunciatory stance *after* doing the research for this book and learning, first hand, how climate science is done in practice (the reader should not be confused by the fact that I start this book with a critique in the introduction). In this sense, I agree with the sociologist Harry Collins (1996: 240), who argues that epistemographers' activism (to which he refers as 'analytical critique of science') grows from, and must be informed by, their scholarly analysis rather than preceding it because 'The thing to notice about analytic critique of science is that, as I have argued above, the commitments which one might finish with are not predictable prior to the analysis'. In this way, the epistemographers' contribution to the practices or community under study cannot be predicted beforehand, as it develops throughout the research as the actors involved negotiate the meaning and future uses of epistemographic analysis (Zuiderent-Jerak, 2015; Box 10).

Knowledge as collective practice

'Epistemography' and 'ethnography' are interrelated endeavours, as the constitution of the community (ethnos) that creates and sustains certain knowledge (episteme) occurs simultaneously. Epistemographers insist that knowledge is not an individual possession, but a collective achievement (Kusch, 2002). Individuals arrive in the world possessing little or nothing in the way of knowledge, and they acquire and develop new knowledge through interactions with relatives, friends and other significant people and institutions implicated throughout the life of a human being, from birth to death (see Berger and Berger, 1972 for an excellent introduction to sociology through the analysis of the main stages of social experience in the biography of individuals). The fact that knowledge only exists and persists as long as it is shared by a group does not mean that epistemographers cannot study the work of single individuals (an epistemography can share aspects of a 'biography'). In fact, the present epistemography is centred around two scientists in particular: Professor Rob Wilson, the principal investigator of a research project called the 'Scottish Pine Project', and Rob's doctoral student, now Dr Miloš Rydval. Epistemographers must keep in mind that scientists and experts who work solo must be sensitive to what colleagues are doing and must coordinate with them if their ideas and skills are to be collectively accepted as expert knowledge (Barnes, 2001: 33). Epistemographers' task is, therefore, to show how knowledge and society are simultaneously created and maintained. In the language employed by many epistemographers, the task is to explain how 'cognitive order' and 'social order' are co-produced (Mazzotti, 2008; Shapin and Schaffer, [1985] 2011). The sociologist Barry Barnes (1985: 82) exemplifies this anti-individualist idea with an example from mathematics:

> If an individual knows Euclid's geometry up to the twentieth theorem we can straightforwardly say that he is in a position to prove the twenty-first theorem: he knows all it is necessary to know. But imagine that this knowledge is spread over the members of a society, some known by some individuals, some by others. We cannot say of this society that it knows enough to prove the twenty-first theorem. To think of the society as individual writ large in this way would be quite misconceived. Suppose that the different individuals, with the different necessary bits of knowledge, did not know each other, or how to find each other. Or suppose they did not trust each other, or know how to check on each other's trustworthiness. In both cases, the twenty-first theorem would remain unproven. The technical knowledge would have been present in the society, but not the necessary internal ordering – the necessary social relationships – for the proof to be executed. Individuals would have known enough mathematics, but not enough about themselves.

Epistemographers have typically focused on the study of historical and contemporary 'practices' as a way to have empirical access to the formation of knowledge. In this respect, the 'Actor-Network Theory' approach (Michael, 2017; Latour, 2005) and the literature on 'praxiography' (Bea, 2017; Mol, 2002) offer two important insights for the study of practices. First, an emphasis on the generative aspects of knowledge and representational practices – in the sense that practices 'enact' stable associations of actors (Law, 2008) or 'perform' social and cognitive order (MacKenzie et al., 2007). Secondly, an insistence on the decentralisation of the knowing subject (Mialet, 2017, 2012) and on the role of bodies, tools and beings in the making of knowledge. As the philosopher and anthropologist Annemarie Mol (2002: 48) explains, 'The knowledge incorporated in practices does not reside in subjects alone, but also in buildings, knives, dyes, desks. And in technologies like patient records.' Consequently, the core of this book consists of five chapters, each being about one 'socio-material' practice involved in the creation of Figure 0.1. In their study of practices, epistemographers should be wary of the perils of both behaviouralism, which consists of focusing on observable behaviour and ignoring the implicit assumptions underlying the enactment of practices and the meanings that people ascribe to them (Geertz, 1973), and idealism, which conceives knowledge as resulting from individual or collective states of mind rather than socio-material configurations and embodiments (Lawrence and Shapin, 1998; Shapin, 1991).

Epistemographers, like ethnographers, have long employed methods of observation and participation to generate data of scientific and technical practices (Hess, 2001). This book is precisely the result of observing the life and work of a group of scientists while being alongside them for four years (as I explain in Box 5, p. 65, this co-presence varied throughout the research). I generated observations and fieldnotes of the work of these scientists by participating in their daily routines, and in this way they came to treat me as one of their own and I acquired a near-native understanding of their collective practices. More specifically, I participated in three fieldwork expeditions (August 2012, August 2013 and August 2015); from April 2012 until April 2013, I also worked one day per week as a voluntary technician for Miloš, Rob's PhD student; in 2013 I audited a course that Rob taught at St Andrews University; I attended a workshop in St Andrews in April 2013, a one-week training course or 'fieldweek' in Tasmania in January 2014, a conference in Melbourne in January 2014 and another one in Aviemore in May 2014. Until December 2014, I visited Miloš at his house almost every week to observe how he worked and I attended all the meetings between Rob and Miloš that took place in St Andrews and in Rob's house in Penicuik. Finally, I accessed Rob and Miloš's presentation slides and article drafts and I have copies of the email interactions where I was 'cc'd'. Unless otherwise stated, all the quotes in this book are from the expeditions, conferences and meetings I attended at the time.

While the enactment of practices always includes both 'doing' and 'thinking' with others, it also involves both routine and change. Practices are repetitive, automated and habitualised patterns of action, but they are also dynamic and mutate over time as people negotiate their correct execution in specific contexts. Precisely, the sociologist Harold Garfinkel (1967: 35) sought to illustrate the ongoing (re)constitution of social norms and practices by devising a series of 'experiments' that disrupted such norms and generated conflict (e.g. a student was told as part of an experiment to distrust the bus driver about the route that would be taken and to record the driver's reaction (Garfinkel, 1967: 35). Epistemographers (myself included, see Box 4, p. 50) have followed a similar methodological strategy and focused on the study of 'scientific controversies' whereby existing practices are challenged and redefined (Collins and Pinch, [2002] 2014; [1993] 2012; 2005). In this sense, while the practices described in this book might seem stable and unproblematic, they are not really so. As I shall show throughout the book, and particularly in the last chapter, these practices have undergone, and are presumably still undergoing, profound changes and have resulted from intense scrutiny from practitioners and outsiders alike.

Formation stories

An epistemography is typically a study of a given 'epistemic object' or representation of knowledge. Experts commonly create maps, diagrams, concepts, theories and numbers, among other objects, to represent previously unknown phenomena through texts and images (Coopmans et al., 2014). Representations abound in our lives, and epistemographers should identify and analyse those privileged epistemic objects towards which experts and the wider society are orientated. The main general question motivating an epistemography is: 'How do certain experts come to know whatever is known about the world, and how do they represent this knowledge and reality?' I myself started out the research that led to the present epistemography by asking a related question: 'How are scientists able to know whether the Earth's temperature has increased over time, and how do they represent historical changes in climate?'

While the epistemographer must defer to subjects any judgement regarding the accuracy of their representations, he or she should consider the extent to which epistemic objects relate to the reality that they are taken to represent (in this sense, an epistemography might share aspects of an 'ontography', as defined by Michael Lynch (2013)). Determining the empirical grounding of epistemic objects is important, not only to give a proper account of knowledge formation but also to dispel charges of 'idealism'. Epistemographers generally espouse a 'materialist' theory of knowledge, which suggests that the environment and the material world surrounding us (including our body and the tools and techniques that render a phenomenon problematic) partake in the knowledge that collectives generate (Monteiro et al., 2012; Østerlie et al., 2012; Bloor, 2008). As Barry Barnes (2013: 119) explains, 'The material world affects us much as it affects animals, through our perceptual apparatus, and thereby modifies our memories and our inductively based inclinations, causally'. Crucially, epistemographers do not assume that epistemic objects are the direct result of the sensory experience of the reality 'out there'. Individuals would not be able to describe and interpret reality in ways collectively judged as competent unless they relied on a tradition of concepts, theories and instruments of observation.

Unlike everyday objects, epistemic objects are never complete and fixed. To an outsider, epistemic objects might look like definitive and solid objects. For instance, Figure 0.1, which was published in 2017, could be seen as the most up-to-date representation of past climate changes in Scotland. However, for the scientists who produced it, Figure 0.1 is already outdated, as they are generating new data and a newer version of such a graph. As the sociologist Karin Knorr Cetina (2008: 89) explains, 'Objects of knowledge appear to have the capacity to unfold indefinitely. They are more like open drawers filled with folders extending indefinitely into the depth of a dark closet. Since epistemic objects are always in the process of being materially defined, they continually acquire new properties and change the ones they have.'

If epistemic objects are seen as 'processes' rather than 'things', the task of epistemographers becomes the study of 'knowledge in action' or the making of knowledge. This anti-essentialist view is rooted in 'process philosophy' more generally (Seibt, 2018; Whitehead, 1929), and on a tradition of social constructivist accounts of 'scientific facts' specifically (Barnes, 2013; Latour and Woolgar, [1979] 1986; Knorr Cetina, 1981; Fleck, [1920] 1981). In his book *Science in Action: How to Follow Scientists and Engineers Through Society*, the philosopher Bruno Latour

(1987) famously equated 'scientific facts' with 'black-boxes', whose inner workings are no longer visible once they are 'black-boxed' and accepted. From Latour's perspective (2005; 1987: 13), epistemographers should describe the constitution of networks of humans and other entities and beings (machines, materials, animals) through which 'scientific facts' achieve the solidity, stability and durability associated with black-boxes:

> We will enter facts and machines while they are in the making; we will carry with us no preconceptions of what constitutes knowledge; we will watch the closure of the black-boxes and be careful to distinguish between two contradictory explanations of this closure, one uttered when it is finished, the other while it is being attempted. This will constitute our *first rule of method* and will make our voyage possible. (Latour, 1987: 13)

Epistemographies should therefore be explanations of the way historical or contemporary epistemic objects temporally 'stabilise' and 'solidify' as convincing accounts and representations of reality. The sociologists Daniel Hirschman and Isaac Ariail Reed (2014: 268) usefully characterise this form of causal explanation as 'formation stories'. They clarify, 'When we speak of *formation stories*, we refer to explanatory accounts of how social kinds are shaped, reshaped, or brought into being; in contrast to forcing causes, these stories take as their points of reference the nonfixedness of social entities, the eventfulness of social life, the emergence of social entities from processes of assemblage, and the dependence of such assemblage and nonfixedness on representation.' Merging description and explanation, as formation stories should do, and transitioning from the narrative of the way people do things to the interpretation of why they do so, are also characteristic of ethnographic reasoning more generally (Katz, 2002, 2001).

A formation story would retrospectively show the sequence of events and practices that have led to an epistemic object. In this way, I suggest, an epistemography should provide a 'wide' and 'slow' description that traces the unfolding of epistemic objects over an extended period of time (see Neyland, 2008: 90 for a discussion of time in ethnographic research, which also applies to issues regarding the 'epistemographic time'). Accordingly, the present epistemography is organised into five chapters, each outlining a stage/practice involved in the creation of Figure 0.1. I have been faithful to the scientists' sense of order (they orientated their work towards the last stage) and I have titled each empirical chapter using their own vocabulary. Chapter 1 is called 'Fieldwork' and discusses the creation of samples from Scots Pine trees during fieldwork expeditions in the Scottish Highlands. Chapter 2 is titled 'Dendrochronology' and describes the production of carefully dated tree-ring data from those wood samples in the form of a long tree-ring chronology. Chapter 3 is titled 'Standardisation', which refers to the creation of tree-ring indices by removing perturbing non-climatic factors from tree-ring chronologies. Chapter 4 sets out the stage of 'Reconstruction' and the establishment of extrapolations of unknown past climates from cleaned tree-ring data. Chapter 5 describes a scientific 'controversy' in which Figure 0.1 was mobilised to refute the hypothesis proposed by one of the parties involved.

To be clear, a formation story is an *artefactual* chronology of the cycle of epistemic work. The stages that I outlined above did not occur linearly, but simultaneously and iteratively. The linear structure of formation stories is, as I see it, a necessary compromise for achieving legibility. Paul Atkinson (1992: 6) explains the dilemma encountered by all social scientists: 'The more readable the account, the more it corresponds to the arbitrary conventions of literary form: the more "faithful" the representation (conventional though it still must be), the less comprehensive it must become.' I have tried to convey a sense of disunity throughout the book by starting each chapter with a presentation of what I call 'epistemic conundrums'. These conundrums are problems that the scientists under study faced at each stage of the process and that they resolved, partly and temporarily, by enacting each specific practice. The resolution of each conundrum involved specific communities and developed in distinct 'epistemic geographies' (Mahony and Hulme, 2018); moving from the field site, the laboratory and the private home of the researcher to the blogosphere, conference rooms and peer-reviewed journals.

While writing a formation story it is important to spell out the broader public relevance of the detailed descriptions and analyses of the epistemic work. Accordingly, in the conclusion sections of each chapter, I have analysed the empirical material in a way that the resulting analysis addresses some of the public concerns about climate science that were formulated after the hacking episode in November 2009 (see preface). In Chapter 1 I conclude that the production of credible samples was simultaneous with the constitution of an expert and exclusive community of fieldworkers who maintained their trust relations through participation and examination of each other's work, year after year, in the rituals of labour and recreation in the field and at home, and I bring to bear this conclusion on public debates about secrecy and openness in climate science. In Chapter 2 I conclude that the creation of an accurately dated tree-ring chronology was possible because the scientists involved initially accepted a mainstream dendrochronological practice (the 'principle of crossdating'), and then critically examined the applicability of such principle to the Scottish context; I draw on this conclusion to respond to public discussions about tribalism, dogmatism and 'groupthink' in climate science. In Chapter 3 I conclude that the removal of non-climatic factors from the tree-ring chronology of Scotland was achieved through the strategic and goal-orientated use of communicative channels in the community under study (laboratory visits and workshops), and in this way I criticise public demands for a 'disinterested' pursuit of climate science that isolates scientists from the social and political context in which they work. In Chapter 4 I conclude that the accuracy of the Scottish reconstruction was established by relying on an accepted precedent and comparing its similarity to canonical temperature reconstructions; I use this conclusion to contribute to public debates about the nature of scientific consensus. Finally, in Chapter 5 I show that a scientific controversy was temporally and partially foreclosed by the mobilisation of a small group of expert colleagues and friends and the participation of the blogosphere, which I see as evidence for the need to reconsider the analytical value of the vernacular label of 'climate sceptic' and to acknowledge the social preconditions of scientifically productive 'civil' scepticism.

Meta-epistemography

Epistemographers should aim to explain the formation of epistemographic knowledge itself. I do not mean that we should communicate our methods and explain how we go about doing our research (why we chose an epistemic object, years of observation, number of interviews, written sources, observation, transcription and analytical practices). With more or less success, such methodological accounts are already part of most social science texts (including the present epistemography) and should continue to be so. I do not propose either that epistemographers include narrative accounts of fieldwork or 'fieldwork tales', even though these texts could become a primary form of evidence (as in Tsing, 1993; Rabinow, [1977] 2007). Instead, what I suggest is that epistemographers should foreground the everyday practices that are often made invisible in accounts of how social scientific knowledge is made (Latimer and López, 2019; Lamont et al., 2011) and interrogate the bases and sources of credibility of their own claims. We should ask: What makes an epistemography (perhaps our own) look credible to subjects, colleagues and other audiences? How does an individual's epistemographic research become part of collectively accepted sociological, anthropological and historiographical knowledge?

A 'meta-epistemography' or an epistemography of epistemography would make epistemographic knowledge amenable to empirical analysis in fundamentally the same way as any other form of knowledge. Epistemographers should spell out the social processes involved in the elaboration and evaluation of an epistemography: the socialisation stages through which epistemographers have been educated in certain academic communities and become part of selected traditions of thought and research practice; the circumstances of production, reception and usage of an epistemography; the interactions established with the subjects and objects implicated by the research; and the distribution of taken-for-granted beliefs and practices according to status and membership of those involved in their research (e.g. 'male or female'; 'PhD student/advisor'; 'outsider/insider'; 'subject/object'; 'publisher /author').

Interrogating the grounds of credibility of epistemographic knowledge is important for two reasons. First, it helps us to understand better systems of expert knowledge. Epistemographers who do not use their own research experience as 'data' are losing an opportunity to refine theories and concepts that explain the constitution of epistemic objects. As I see it, epistemographers should aspire to generate theory that is 'self-exemplifying' (Merton, 1978) or applicable or 'reflexive' to itself (Bloor, [1976] 1991) in order to generate more systematic and explanatory knowledge of our societies. The second reason why we should carry out a meta-epistemography is to make sociological, historiographical and anthropological knowledge more credible. While I agree with Michael Lynch's (2000: 36) warning that '[w]hat reflexivity does, what it threatens to expose, what it reveals and who it empowers depends upon who does it and how they go about it', developing a heightened awareness of the conditions of our mode of knowledge production might give epistemographers an (ad)vantage point for producing more convincing accounts. The reason for this is exactly the same one that I offered in the preface when I justified the importance of explaining the internal workings of climate science. Epistemographers, like any other expert, are expected by the broader public who often fund our research to provide evidence of our trustworthiness; hence, we should spell out our collective forms of reasoning and validation of epistemographic evidence. As the sociologist Michael Duneier (2011: 10) suggests:

> We can best improve our methods by engaging in practices that reassure our readers that they can trust they know *how* they have been convinced. It is a lack of transparency that results in a sense that the wool is being pulled over a reader's eyes. Our goal should be to institutionalize methods that make it normative for us to be as up front as possible about how we have achieved our effects.

My contribution to the institutionalisation of epistemographic reflexivity and increased public awareness of the making of social scientific knowledge is exemplified in this book with an account of the development of the present epistemography. In doing so, I was inspired by Malcolm Ashmore's (1989) strategy of using a double-text and Annemarie Mol's (2002) strategy of spreading out a subtext on the bottom of the pages of her book reflecting on the relationship between her research and various scholarly traditions in sociology, philosophy and anthropology. Unlike Mol, however, I shall specify the connections, both in time and in content, between the main text (the epistemography of climate science) and the subtext (the context and process of production of such epistemographic analysis) by offering a series of numbered boxes of 'epistemographic vignettes' throughout the book.

Conclusion

This chapter has developed the concept of 'epistemography' as a shorthand for referring to a set of standard methodological practices in the STS field aimed at researching and writing about technical and scientific objects and systems of expertise of relevance in contemporary and past societies. I have also used the presentation of, what I suggest are, the four main rules of method establishing the appropriate conduct of sociological, anthropological and historical epistemographies to introduce the features of the present epistemography of climate change.

The first principle of 'situated impartiality' requires epistemographers to refrain from resolving disputes about what 'knowledge' is and, instead, offer socio-historical explanations for the ways certain claims are simply accepted as 'known', which are sensitive to the working life of the epistemic community studied. This is why this book is an epistemography *of* climate change as opposed to *in* climate change: I have neither the interest nor the competence to tell climate scientists what they should know or how they should go about studying climate change. As an ordinary citizen, I defer these questions to climate scientists and I accept whatever they tell me as truthful. Instead, my aim is to tell interested audiences how a group of climate scientists have come to know about climate change. To do so, I have explored the scientific community in question from the inside, unafraid to get involved in their working lives and struggles for public credibility. Unlike traditional views of scientific impartiality that privilege detachment from the object/subjects of study as a requirement for knowing, the notion of 'situated impartiality' regards the unexpected interventions and contributions that derive from the practice of social research to be normal constituents of social scientific knowledge.

The second principle stipulates that the epistemographer must regard 'knowledge as collective practice' and generate first-hand observations or historical evidence of the way experts conduct their work in alignment with their relevant socio-natural environment. This book has resulted from conducting a three-year-long participant observation of the work of a small group of dendroclimatologists (Professor Rob Wilson and the team participating in the Scottish Pine Project, particularly his then PhD student Miloš Rydval) who wanted to learn how Scotland's climate has changed throughout the last 2,000 years from the analysis of Scots pine tree growth. In my analysis of the socio-material practices of dendroclimatology, I shall seek to ascertain the extent to which Figure 0.1 results from the specific arrangements and features of the academic community to which Rob and Miloš belonged and the Scottish environment in which they worked.

The third principle of providing 'formation stories' requires the provision of explanatory descriptions of the coming into existence of representations of knowledge. The present epistemography should not be read as tracing the development of a pre-existing epistemic object (for a retrospective analysis of 'climate knowledge in action' see Howe, 2014, as it outlines the historical constitution of the 'Keeling curve' in the second half of the twentieth century used to monitor the concentration of carbon dioxide in the Earth's atmosphere). Instead, this epistemography results from a synchronic investigation of the creation of Figure 0.1. As I believe that epistemographers should strategically conduct analyses of popular and widely accepted epistemic objects, formation stories should be written for a wide readership. In this way, in each chapter I have sought to connect the specific local problems and 'epistemological conundrums' experienced by Rob and Miloš (in Wright Mills's terms 'private troubles') to broader concerns and social problems in contemporary Western societies (in Wright Mills's terms 'public issues'; Box 10).

The fourth principle, an 'epistemography of epistemography', specifies the need to write a formation story about epistemographic knowledge in order to further the scholarly understanding of knowledge-making processes and the public credibility of epistemographers. The links between the two epistemographies in this book – about the making of scientific knowledge of climate change and about the making of sociological knowledge of climate science – are illustrated within each chapter through non-lineal analyses of research events (Boxes).

Besides introducing the present epistemography of climate change, the main purpose of this chapter has been to provide the interested reader with the necessary tools for expanding the range of objects of epistemographic study to include new forms of expertise proliferating in our increasingly specialised societies. We are surrounded by 'analysts', 'specialists' and 'experts' who influence the way we live our lives through their representations. A few examples of existing epistemographies of ubiquitous epistemic objects are: 'nutrition labels' that hugely affect food choices and health habits (Frohlich, 2017); the 'Body Mass Index' used to measure body fat and establish the extent of the obesity pandemic (Fletcher, 2014); 'fire safety regulation and evacuation plans' prevalent in all public and private buildings (Spinardi et al., 2017); the 'Mindfulness-Based Stress Reduction Programme' very popular in business circles (Arat, 2017; Braun, 2017); 'wine-tasting quality standards' used as formalised methodologies for establishing expensive classifications of wine (Phillips, 2016; Shapin, 2016); the 'Gartner's Magic Quadrant' widely employed in market research (Pollock and Williams, 2016); the 'Gaussian copula model' used to price financial products (MacKenzie and Spears, 2014a, 2014b); and 'DNA fingerprinting', one of the most important forensic techniques in criminal investigations (Lynch, 2008).

As I see it, epistemographers should prioritise the analysis of epistemic objects that, like those listed above, are used for structuring markets and governing societies globally (Miller, 2007). I envisage new epistemographies of global epistemic objects whose local constitution and deployment, as far as I know, remain unexplored. New epistemographies could examine the 'Strengths, Weaknesses, Opportunities, and Threats (SWOT) matrix', a very popular strategic planning technique; the 'Myers–Briggs Type Indicator Personality Test', a human resources test widely used for selecting candidates in many public and private sector institutions; the 'GROW model of coaching' that has gained considerable influence in the corporate world; the 'terrorist threat levels' globally used by governments to inform on the terrorist danger facing their nations; and the 'Programme for International Student Assessment (PISA)' intended to evaluate educational systems worldwide. The possibilities of epistemography are endless because the creativity of humans and their desire for knowledge and power have no limit (Foucault, 1980).

1

Fieldwork

Every year at the beginning of August, since 2006, Professor Rob Wilson has been busy putting the finishing touches to an annual fieldwork expedition in the Scottish Highlands. Rob is the leader of the 'Scottish Pine Project', a dendroclimatological project aiming to use Scots pine trees (*Pinus Sylvestris L.*) to reconstruct the climatic history of Scotland over the last two millennia. During fieldwork, the members of the Scottish Pine Project and other occasional participants like me collect pieces of Scots pine wood from forests, buildings and lakes across the Scottish Highlands. Rob and his colleagues referred to these pieces of wood as 'samples'. To count as a sample in the Scottish Pine Project and in most dendroclimatological projects, the wood has to yield useful information about changes in temperature or precipitation from year to year, as reflected in the variation in the width of the layers of tree growth or so-called 'tree-rings'. Samples that have variable patterns of tree-rings are described as being 'sensitive' to climate, as opposed to 'complacent' samples that show a uniform sequence of growth.

Whether a sample is a good source of climatic information depends on the tree species and its location. Trees growing in extreme environments are known to have more variable tree-ring patterns, because changes in weather (more rain or warmer temperatures) affect their growth more distinctively. In Alpine climates like the Scottish Highlands, trees are incapable of growing because it is too cold and wet. The authors of one of the first dendroclimatology textbooks, Marvin Stokes and Terah Smiley (1968), illustrated the relationship between growing conditions and tree-ring patterns with a drawing of one tree growing on water-saturated ground (which produced 'complacent' tree-rings) and another tree on a rocky, dry slope (which produced 'sensitive' tree-rings). Dendroclimatologists commonly sample those areas or 'sites', like the rocky slopes of the Scottish Highlands (Figure 1.1), where trees are most likely to be affected by changes in precipitation or temperature and hence produce sensitive tree-rings. This sampling practice is referred to as the 'principle of site selection', which has the purpose, as one textbook author explains (Speer, 2010: 21), 'to maximize the [climate] signal recorded in the trees'.

Initially, the Scottish Pine Project team only sampled Caledonian forests growing on the high-elevation areas of the Scottish Highlands because Rob knew that Scots

1.1 The fieldwork practice of purposively sampling trees that grow on rocky slopes ('the principle of site location') is based on the assumption that their rings will provide better climatic information because they are more 'sensitive' to changes in climate.

pine trees growing at such locations were very sensitive to changes in temperature. In 1984, the dendroclimatologist Malcolm Hughes published the first temperature reconstruction for Scotland from this tree species (Hughes et al., 1984); thirty years later, Rob sought to build upon Hughes's work by expanding the number of samples and their geographical extension. Building on previous studies of past ecologies, Rob had learnt that, due to deforestation, only about 1 per cent of the ancient pine woodlands in the Scottish Highlands remains. Rob's aim was to conduct an exhaustive sampling of this 1 per cent of remnant pines. As Rob wrote, optimistically, on the website of the Scottish Pine Project, 'Long term plan is to sample ALL remaining semi-natural pine woodlands in Scotland. We are almost there' (Wilson, 2019).

In order to generate more samples from places other than pinewood forests, Rob assembled a team of colleagues who helped him to identify and extract partially fossilised wood preserved in lakes (which they called 'subfossil') and historical buildings across the Scottish Highlands. With these three types of samples (living, subfossil and historical), they had built up what they called a 'network of sites'. In May 2014, Rob sent me an email with two maps saying, 'This is where we were in May 2006 and this is where we are now' (Figure 1.2). One map had only seven dots, representing the areas that Hughes had sampled in the 1980s. Rob used these

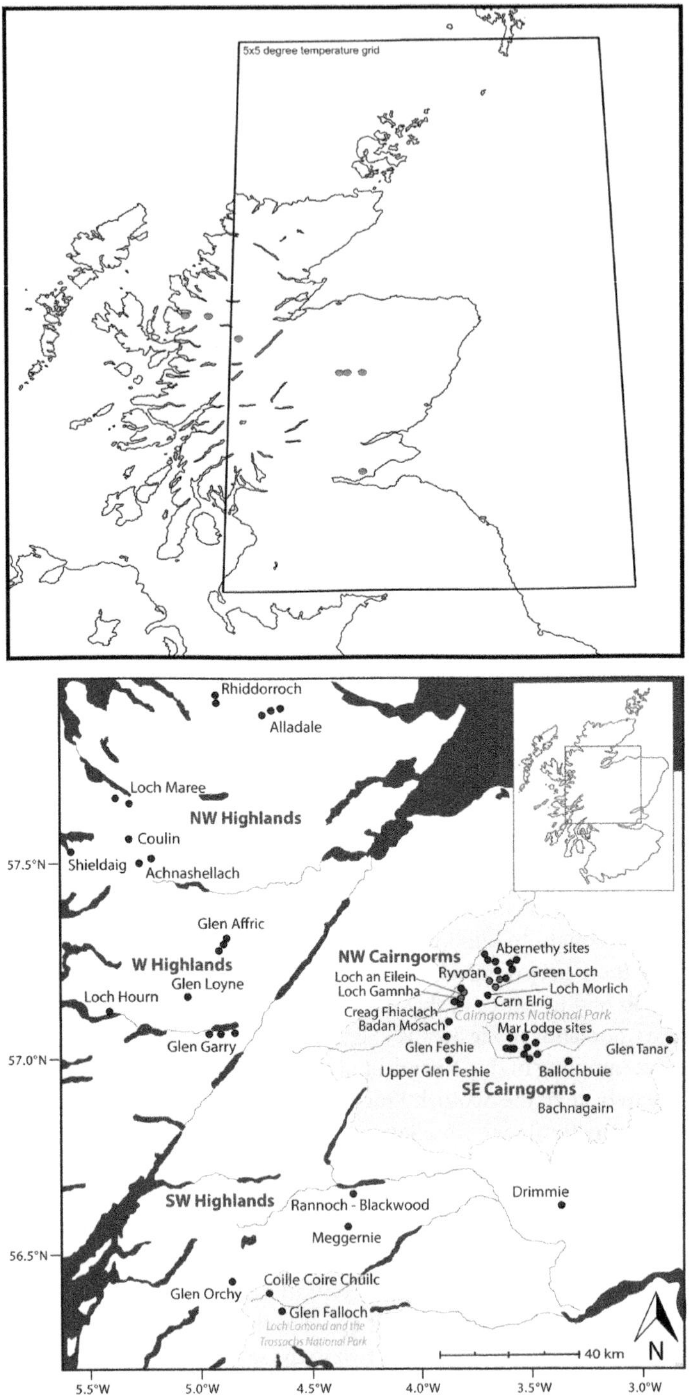

1.2 Rob and Miloš created these two maps to illustrate the growth in the number and geographical scope of the sampling sites since the start of the Scottish Pine Project in 2006.

sites as a starting point in 2006, when he initiated the Scottish Pine Project. The other map had forty-four dots, and included the names of all the new areas that Rob and his team had sampled up to 2014. This second map has kept growing since I finished the research for this book, as Rob and his team have continued sampling new areas and generating new data, which is why Figure 0.1 should be seen as an incomplete epistemic object (see Introduction).

The sampling design of the Scottish Pine Project was not only purposive and geographically extensive but also iterative. The three expeditions in which I participated (2012, 2013 and 2015) were concentrated around two lakes in the Cairngorms – Loch an Eilein and Loch Gamhna – in the eastern part of the Scottish Highlands. Every year, we sampled a few metres of the lake banks. Likewise, we always extracted two or three wooden cores from each living tree and a couple of subfossil disks from each submerged log. Rob defended the decision to take multiple samples by saying that he was a 'great believer in the principle of replication'. This principle states that the climate signal and information that is assumed to exist in trees can be enhanced by averaging the data from replicate samples. Rob and his dendroclimatology colleagues believe that the 'noisy' factors affecting tree growth will be minimised if data from more and more samples are averaged (Chapter 3). The fieldwork practice of generating as many samples as possible explains why Rob welcomed students, partners and amateurs like me who, if properly trained, could become useful workforce.

Dendroclimatologists have historically been criticised for their sampling strategy by those who uphold random sampling as the standard of scientificity as it supposedly eliminates the risk that samples reflect the researcher's point of view. In his textbook, Harold Fritts (1976: 17) dismissed 'Critics in the past who have questioned the validity of selecting the most drought-sensitive trees for sampling rather than selecting trees randomly (Glock, 1955). Such judgement fails to recognise that the dendrochronologist has a particular strategy in mind which requires that his samples be affected similarly by a given set of growth-limiting factors.' In one interview I conducted with a reputed dendroclimatologist, he told me about an 'incident' that happened to him in the early stages of his career in a conference talk, when a member of the audience criticised his results for not 'having degrees of freedom'. This dendroclimatologist explained to me, 'You see? This is the type of pure statistician response as classically considered with an experiment design with random sampling and control; that was the way she was educated. But if you are doing a climate reconstruction, and time matters, we must try to select the oldest and most sensitive trees. There's no way around it!' When I once asked Rob his opinion about this matter, he was equally convinced that purposive sampling in dendroclimatology was justified. He expressed his opinion with an eloquent analogy: 'You would not go to the tropics to study glaciers!'

While Rob and the other dendroclimatologists to whom I talked were convinced that purposive sampling was the most appropriate strategy for dendroclimatological projects ('There's no way around it'), they have also been examining the possibility that their sampling strategy has some limitations. In particular, a group of dendrochronologists have recently identified the 'Modern-Sample Bias', which is seen to arise from sampling only the oldest trees (Melvin et al., 2013). Similarly, at an international dendrochronology conference I attended in Melbourne in January 2014, one of the most commented on presentations among the attendees to whom I talked was about the biases and effects of different sampling strategies on the resulting data (this presentation was eventually published in a paper; Nehrbass-Ahles et al., 2014).

The 'epistemological conundrum' that Rob and the members of the Scottish Pine Project faced at this stage of the making of Figure 0.1 was to satisfy themselves and their critics that, despite the known limitations of purposive sampling, the samples they had generated during fieldwork were untainted by personal or systematic bias and were a reliable starting point for producing knowledge of past climate in Scotland. In this chapter, I explain how they sought to resolve this challenge by detailing the work involved in producing credible samples; such work is narrated around a typical day of fieldwork in the Scottish Pine Project expeditions that I attended.

Labour of fieldwork divisions

The setting up and running of the Scottish Pine Project fieldwork expedition depended on substantial coordination in establishing divisions of expertise, responsibilities and tasks, and distributing the appropriate materials among participants. The philosopher Rolland Munro (1997) coined the expression 'labour of division' – as distinct from the traditional notion of 'division of labour' – to refer to the work of creating and upholding divisions that characterise the functioning of complex organisations and societies, including scientific ones. As the leader of the fieldwork expedition, Rob was in charge of arranging people in terms of their associated duties and skills, and the success of the expedition essentially lay in fieldworkers' accepting such arrangements.

The labour of divisions involved in the Scottish Pine Project fieldwork expedition was clearly reflected in the timetable that Rob sent to participants in the weeks leading up to the expedition in August 2013, where we were distributed by days, sub-teams and 'number of beds needed' (Figure 1.3). Timetabling took more than six months of preparation: we needed to agree on the exact dates of the expedition; Rob had to apply for and secure funding to cover the trip's expenses; he had to prepare and check all the necessary equipment, request access to the sampling areas from landowners and from government agencies, and fill in the safety and insurance forms; and he had to find accommodation and transportation for all fieldworkers.

Rob created sub-teams that worked independently of each other, each under the supervision of a different leader. 'Stewart' was responsible for the 'lake sonar survey team', while 'Leah' was in charge of the 'historical sampling team' with 'Anne', Rob's technician (their identities have been anonymised). I was in the 'lake subfossil sampling' and the '10mm tree coring' teams led by Rob and his friend Dr Björn Gunnarson, the head of a dendroclimatology laboratory in Sweden. Rob's undergraduate and doctoral students were also part of the 'lake subfossil sampling'

Participants	Wednesday 28/08/2013	Thursday 29/08/2013	Friday 30/08/2013	Saturday 31/08/2013	Sunday 01/09/2013	Monday 02/09/2013	Tuesday 03/09/2013	Wednesday 04/09/2013	Thursday 05/09/2013	Friday 06/09/2013	Saturday 07/09/2013	Sunday 08/09/2013	Monday 09/09/2013	Tuesday 10/09/2013
Rob Wilson		X	X	X	X	X	X	X	X	X				
[redacted]				X	X	X	X	X	X	X	X	X	X	
[redacted]				X	X	X				X	X	X	X	
[redacted]	?	X												
[redacted]	X	X												
[redacted]	X	X												
[redacted]	X	X												
Milos Rydval					X	X	X	X	X	X				
Bjiorn Gunnarson					X	X	X	X	X	X				
Hans Linderholm					X	X	X	X	X	train				
Meritxell Ramirez Olle							X	X	X	X				
[redacted]		X	X	X	X	X	X	X	X	train				
[redacted]					X	X	X							
No. of beds needed	3	6	2	4	8	8	8	8	8	6	2	2	2	0

X	night in aviemore
	Driving up
	Driving back
	Sonar survey work
	Historical sampling
	Lake scouting
	10mm tree coring
	Lake sub-fossil sampling

1.3 As seen in this timetable, fieldwork required substantive labour of division of time, people, tasks and equipment (including beds and cars). For those wanting to remain anonymous, I have blanked out their names.

and the '10mm tree coring' teams. Notably, Miloš had been Rob's undergraduate student at St Andrews and, in Rob's words, was 'one of the best undergraduate students I have ever supervised'. After Miloš graduated in 2008, Rob hired him as a technician, and later, in 2011, supported him in a (successful) application for a doctoral scholarship that allowed Miloš to produce Figure 0.1 as part of his thesis. Björn's Master's student, 'Emily' was also part of these two teams in the expedition of 2013.

Rob's ability as the leader and coordinator of the Scottish Pine Project fieldwork expedition lay in making sure that fieldworkers contributed to the common goal of producing samples while they also achieved their own respective interests. Rob described to me his role as 'making sure that nobody steps onto someone's shoes'. Each fieldworker had a specific interest in producing samples: Stewart wanted to find subfossil wood in order to refine his sonar survey system; Leah sought to use the archaeological wood to continue her exploration of the cultural heritage of Scotland; Björn hoped to use the Scottish data to verify his climate reconstruction for Scandinavia and for other research purposes; and Miloš and I wanted to complete our respective PhD theses (Box 1).

Box 1

Why doing fieldwork about fieldwork mattered

Joining the fieldwork expedition in the Scottish Highlands early on in my research was, retrospectively, the best decision I ever made. Rob asked me if I wanted to join the expedition in May 2012, a month after I had agreed with him and Miloš that I would conduct my study while working for them as a voluntary technician. I accepted Rob's invitation because I sensed that doing fieldwork was important to him. 'All the [Scottish Pine Project] team is going to be there, and you'll meet them. You'll have fun!' Rob assured me. I certainly had plenty of fun on the three expeditions (2012, 2013 and 2015) in which I took part. However, joining the fieldwork expedition was much more than an enjoyable experience. Doing (ethnographic) fieldwork about (dendroclimatological) fieldwork meant that I not only *observed* how Rob and his team produced samples as an outsider but also *participated* in their production as an insider. My involvement was possible because of the nature of dendroclimatological fieldwork expeditions. Unlike that in other paleoclimatological disciplines, such as ice-core analysis in Antarctica (Skrydstrup, 2012), the equipment used to sample trees is relatively simple and affordable, which facilitates the training and participation of amateurs like me. The first day Rob taught me how to use the sampling equipment, I felt relieved and said to myself, 'I think I can do it.'

As doing fieldwork is part of the 'rite of passage' into the dendroclimatology profession, Rob and colleagues came to see me as if I were one of their own. They sometimes referred to me as a 'dendro-sociologist' and Rob often introduced me to his colleagues as 'a sociologist who is studying us and has done fieldwork with us'. On the website of his laboratory, Rob included me as one of his 'students' and emphasised that I was 'helping both with lab and fieldwork'. Being part of the fieldwork expedition also gave Rob and his team the necessary time and face-to-face clues (Goffman, 1967) to get to know me and to trust me. During the hiking trips, the long sampling hours around the lakes and the sharing of the evening meal, Rob and his team asked me plenty of questions about my work such as: 'So, coming on fieldwork with us is your way of doing fieldwork?' We became aware that we shared many research methods and ideals; not least we were united in the defence of fieldwork as the best method for generating scientific knowledge of the world, whether dendroclimatological or sociological.

Doing fieldwork about fieldwork also made me realise that my own friendly relations with Rob and Miloš, as well as the friendships they maintained

1.4 Friendship as a scientific method.

with their colleagues, were generative of scientific knowledge, both social and natural (Ramírez-i-Ollé, 2019a). I first saw this point more clearly when Rob circulated the fieldwork pictures at the end of my first expedition. There were many pictures where I appeared as an integrated and happy member of the expedition, sometimes taking notes for my own epistemography, and other times taking notes for them and sampling trees. There is one picture (the one included here) where I appear in the middle, Rob on the left of the picture and Miloš on the right, that symbolises the mutuality that formed between us: their interests and work came to matter to me, and, as a result, my research and concerns also became important to them.

The moral economy of fieldwork

The production of samples for the Scottish Pine Project, unlike a market economy, was not an explicit and anonymous exchange of samples for money or another commodity; instead, it resembled a 'gift-giving' exchange between peers who shared the implicit understanding that their individual efforts during fieldwork would be rewarded and reciprocated later on (Hagstrom, 1982; Knorr Cetina, 1982). The historians of science Robert Kohler (1991) and Lorraine Daston (1995) proposed the concept of 'moral economy' to describe the informal, honorific and tacit rules governing academic science (the rules of conduct in 'applied' or 'industrial' science might not be distinctively different; see Shapin, 2008; Dennis, 1987).

In the moral economy of the Scottish Pine Project fieldwork, participants produced samples in exchange for trust and friendship, which, in turn, were used to gain access to research networks, data, metadata and information on the basis of a series of tacit rules over what constituted good and bad behaviour in the field site. Rob implicitly referred to the morality of these exchanges when I asked him to explain his decision of sharing data with colleagues prior to publishing them into a public archive of tree-ring data (the so-called 'International Tree-Ring Data Bank' or ITRDB). Rob responded, 'Basically, I share the [pre-publication] data with those who deserve them.' As the leader of the expedition, Rob was in charge of defining the rules governing the moral economy of fieldwork and the exchange of samples. These exchanges differed in accordance with the three distinct groups involved in the expedition.

With dendroclimatology colleagues who were interested in the climatic value of samples, Rob employed 'priority access to data' as a form of reward for their participation in fieldwork. This idea became clear to me when I asked Rob why Dr Hans Linderholm – Björn's friend and the head of another Swedish dendroclimatology laboratory – was taking part in the 2013 expedition. Rob said, 'Hans is interested in using the Scottish data to study the Summer North Atlantic Oscillation, a large-scale pattern of climate variability.' Rob also told me that he had to be 'fair' to another researcher who had similar interests to Hans, and who had also done fieldwork a few years before him. Prompted by my questioning, Rob specified that 'being fair' in this context meant giving simultaneous access to the data to both Hans and the other researcher. Rob also clarified, however, that he would wait for Miloš to publish articles from his thesis before 'releasing' the Scottish data to the wider scientific community. As Miloš was the main person in charge of generating data from samples (Chapter 2), Rob offered him 'temporary monopoly rights' and the possibility of exploiting the data prior to allowing other senior colleagues to use them.

Money also played an important role in regulating the interactions among Rob and his colleagues. The expedition was like a 'financial union', in the sense that it was supported by a common budget. At the time of my participation, Rob had obtained two different grants (from a private foundation, The Carnegie Trust, and from a public research body, the UK's Natural Environment Research Council) to cover fieldwork costs, Anne's (a part-time technician) salary and Miloš's doctoral stipend and fees. While Rob did not pay fieldworkers for their work, he used the 'fieldwork budget' to cover the costs of food and accommodation for everybody, including me. Rob also paid the travel costs of the official partners in the funding application (Hans, Björn and Miloš) but not for Björn's PhD student or for me; hence establishing a distinction between 'core' and 'occasional' fieldworkers.

Rob also had a reward system in place for undergraduate students who participated in fieldwork expeditions in order to complete their dissertations. Rob assigned and supervised dissertation projects that were related to the Scottish Pine Project. Rob insisted that all his students should do fieldwork: 'It is important that they see for themselves where the data they'll generate comes from.' Rob rewarded students who had done particularly good dissertations with article co-authorships. One of these students told me that she was pleased by 'Rob's generosity'. The fact that this student qualified Rob's gesture as 'generous' indicates that being a co-author was beyond her expectations. For another of Rob's students, becoming involved in the production of samples during fieldwork was an incentive to work harder. 'If it hadn't been for the effort I put in collecting these samples,' she said, 'I think I would have lost interest in the project altogether.'

Finally, Rob established a system of exchange with forestry officials and landowners, who allowed the team to have access to the protected areas and private estates where the selected sampling sites were located. In the three expeditions in which I participated, we worked around the forests and lakes in Rothiemurchus, an Estate in the Northern Cairngorms located five kilometres south of Aviemore. In exchange for unrestricted access to this area, Rob wrote an annual report on the conditions of the woodlands and participated in talks organised by the Rothiemurchus Estate. The relationship of courtesy that Rob had established with the landowners endowed him with a sense of responsibility and ownership towards the sampling sites. In one of the many conversations we had while walking towards the sampling site, Rob told me about a group of dendroclimatologists who had gone on a fieldwork expedition to a foreign country and did not contact the local scientific team. 'I would have felt offended if other dendroclimatologists did the same in Scotland,' Rob noted. 'While everybody is free to come and leave Scotland, I think it would be discourteous not to tell me anything.'

The ethos of the heroic fieldworker

Fieldwork involved hard physical labour, and embracing the repetitive and demanding conditions of producing samples in the field was valued as a professional virtue among participants in the Scottish Pine Project fieldwork expedition, particularly by Rob. What I call the 'ethos of the heroic fieldworker' – after Bruce Hevly's (1996) description of nineteenth-century glaciology expeditions as 'heroic science' – describes the spirit of adventure, direct experience of nature and personal sacrifices that I observed and learned from Rob and his team. Unlike nineteenth-century glacier physics expeditions – which, as Hevly points out, were 'gendered' in the sense that women were excluded from expeditions, and ideas about appropriate conduct in the field were essentially related to manly experience – gender equality was a characteristic of the Scottish Pine Project expedition. The two female dendrochronologists in the team carried out the same arduous activities as their male counterparts and conformed to the same values of effort and sacrifice; in turn, the men routinely executed the domestic labour of fieldwork (shopping, cooking and washing) typically associated with women.

I quickly learnt that it was important to bear the hardships of fieldwork in high spirits during my first breakfast with the team. I was about to eat the usual low-calorie breakfast I had every morning, when Rob shouted at me, 'Hey, I don't want your supervisors to accuse me of mistreating you. Eat something more substantial or your energies will drain in the field.' He offered me porridge, bread, chocolate, jam and biscuits. After breakfast, we prepared a lunch box with a couple of sandwiches and cereal bars to ingest as fast as possible in the forest while avoiding being bitten by the Highland midges (small flies that are characteristic of the Scottish Highlands from late spring to late summer). These insects made our fieldwork very unpleasant. We had to wear nets to avoid their bites on our faces, but doing so was uncomfortable and reduced our field of vision. While the weather was mostly pleasant (this is why Rob scheduled the expedition for late summer, when the Scottish weather is often milder), it sometimes rained, making fieldwork even more miserable.

Adverse weather conditions and midges were only collateral elements of the main strenuous activities of the day: getting to the site; extracting the wood from the trees or logs; and returning the samples, boxes and equipment to the pickup truck. On my first day of fieldwork – as I sat comfortably in the car that Rob was driving to Loch Gamhna and Loch an Eilein on the Rothiemurchus Estate – I learned that we were lucky to have car access to the lakes. Before Rob had established good relations with the estate owners, the team had to walk a minimum of one hour a day to get to the site and then had to carry back the heavy equipment and samples. In the three expeditions in which I took part, we did not walk more than ten minutes from the place where we parked the pickup truck.

Sampling living trees growing on high elevation slopes required some trekking skills, and those in the team who struggled to hike up mountains often joked that 'they should have gone to the gym' prior to fieldwork. Sampling living trees was easier than sampling subfossil trees in lakes because the samples and the equipment were lighter. The equipment needed for sampling living trees was: a 'corer' used for making a hole in the tree and extracting cores of wood; masking tape used to attach a label to a plastic straw; wide plastic straws in which cores were stored; and a marker to write the sample number and code. The corer is the most idiosyncratic tool of fieldwork, and it is also used symbolically as a reward (Box 2). As the corer is inserted into the tree, it becomes harder to turn the handle (Figure 1.5a). Because of Rob's insistence that samples should be replicated, we repeated this procedure twice for each tree for a minimum of twenty trees per sampling site. As a result, our arms felt quite sore at the end of the day.

Box 2

Receiving a 'prize' from subjects: thank you, but why?

I realised that the corer was a symbol for dendroclimatologists when they gave me one as a 'prize' for a talk I gave in January 2014 at an international dendrochronology conference in Melbourne. I came up with the idea of submitting an abstract to this conference when I was trying to justify (to founders and to my supervisors) the time and money spent travelling to the other side of the world to attend the conference where Rob and Miloš had planned to present Figure 0.1 for the first time. After two years of fieldwork, I was also very keen to share some of my ideas with subjects. In my conference talk I essentially argued that trust among dendroclimatologists remained important despite all the technical developments that had occurred. In my own words to them: 'Trust still serves as the social glue needed for your work.' From my conversations with Rob and Miloš, I expected that my talk would provoke some discussion. After a stimulating and intense Q&A, I had the sense that the audience had largely accepted my argument, but I never imagined that the scientific committee of the conference would be so convinced as to give me a prize.

I was both pleased and concerned about receiving a prize from my subjects. I interpreted the prize as a welcoming gesture to their community, and I was obviously very happy and grateful. However, I was also worried because I had conceived my talk as an opportunity to do what the sociologist Pierre Bourdieu referred to as an 'epistemological rupture'. Bourdieu argued that the quality of sociological ideas should be evaluated in accordance with their degree of 'rupture with modes of thinking, concepts, and methods that have every appearance of *common sense*, of ordinary sense, and of good scientific sense' (Bourdieu and Wacquant, 1992: 251). While I did not want to offend my audience, I hoped that my talk would be perceived as provocative and provide some counter-intuitive ideas about the making of dendroclimatology. Using Bourdieu and Wacquant's own words (1992: 251), I sought 'to produce, if not a "new person", then at least "a new gaze", a *sociological eye*' into dendroclimatologists. After accepting the prize, I wondered: 'Is the act of receiving a prize from subjects a "bad sign" in the sense that I have

1.5a

1.5b

1.5c

1.5d

1.5e

1.5 Generating samples by coring living trees with a corer and by cutting wood slices of submerged logs was a physically hard and exhausting team effort.

confirmed their common-sense rather than challenged it?' 'Have I failed to produce a "new gaze", a "sociological eye" into my subjects, and to convert them, if you wish, into "dendro-sociologists"?' After the award ceremony, I emailed my supervisors and shared my puzzlement: 'I am not sure what this prize means.'

As I did not know how to interpret their 'gift', I sought to resolve my doubts in the only way I could think of; which was approaching problems and doubts as empirical questions, and asking two members of the scientific committee why they had given me the prize. One of them said that my talk was 'refreshing', which I took as an indication that it had produced some of the 'epistemological rupture' I had hoped to achieve. The other one said, 'We appreciated that you had the courage to present your work in front of us.' It seemed that the committee valued the fact that I had created an opportunity to discuss my work with their community and the prize was a way of reciprocating my gesture. After this first experience, I gave three more talks at dendrochronology conferences because I understood that this form of interaction was in line with the code of behaviour of my subjects and allowed me to discuss and refine my ideas and potentially challenge some stereotypes without hurting their feelings and being perceived as offensive.

Dragging and extracting pieces of submerged wood out of the water was a much more time-consuming and elaborate activity than extracting pieces of wood from living trees. Sampling one forest could take us half a day, while sampling the banks of Loch Gamhna and Loch an Eilein had taken Rob and his team three years. Sampling submerged logs required a huge team effort. Rob was often inside the lake scouting in search of submerged wood with snorkels and masks (Figure 1.5b–c). Once he identified a tree, he fastened the submerged log with a grabber, which was in turn tied to a rope. Those of us waiting on the banks pulled the log out of the water with a winch and a pulley. Björn gave instructions to Rob, Hans, Miloš and myself on how to position the log so that it was safer for him to cut a slice of wood with the chainsaw. Chain-sawing was seen as the riskiest activity of all and the reason why Rob had bought 'insurance'. He often finished the day by joking, 'Another day and no deaths!'

Despite all the hardships associated with fieldwork, we all valued the fact that the walks to and from the sampling sites and the time spent together sampling trees were an opportunity to socialise and to have fun together in natural surroundings that we all agreed were stunning. Stereotyped comments about the Scottish scenery and history were commonly heard during these walks, especially when Rob mentioned that the seventeenth-century Scottish literary hero, Rob Roy McGregor, lived somewhere near Loch an Eilein where we did most of our lake sampling. Rob and Björn also commonly used these walks to tell us aspects of the environmental history of the areas where we worked or where they had previously worked. To me, these daily walks and conversations did not feel very different from the ones I thoroughly enjoy with friends.

The calibrated body of the fieldworker

Fieldwork was a very stimulating sensory activity that depended on the bodily perception of objects and their surroundings. Like any other instrument, the body of the fieldworker needed to be 'calibrated' and adjusted to communal standards of fieldwork practice (for an extended discussion about the body as a scientific instrument, see Golinski, 1998: 183–185). The 'calibrated body of the fieldworker' was constituted during the daily sampling routines and walks to and from the sites, when Rob and the other fieldworkers trained us to sample and offered snippets of information about the area, the layout of the forest and the characteristics of the trees that surrounded us. When one day in the field site I asked Rob to articulate the importance of doing fieldwork, he succinctly said, 'We've been in the sites, we've done the data, and we know them so well.' The connection that Rob established between the verbs 'being', 'doing' and 'knowing' is crucial to understanding how the active body of the fieldworker, and more specifically the senses of sight, touch and smell, if properly calibrated, were employed to produce knowledge.

Crucial to the production of samples was the visual identification of the relevant tree species; this was a skill that experienced fieldworkers deployed automatically, but it had to be explicitly formulated to neophytes like me. On my first fieldwork expedition, Rob showed me how to identify what trees to sample. As I recall from my field notes, I was standing under the crown of a tall tree, swatting midges, while Rob was getting all the necessary equipment ready:

> 'This is [pointing to a tree] a Scots pine tree, which is the tree species that we want. You don't need to worry about the tree species, because I will bring you to forests that only have Scots pine trees. But in case of doubt, check two things: the leaves are like needles [he gets one needle from the ground and shows it to me]. And secondly, look for pine cones [he gets one cone from the ground and shows it to me]. Then, the next thing you should do is to place the corer into the tree at breast height. Do not sample the tree near the base, because the rings at that level are less concentric and are not representative of the actual growth of the tree.'
>
> [Rob inserts the corer into the tree, turning the handle clockwise. As the friction with the wood increases, we can hear a louder noise as if the tree is crying. Once Rob has inserted more than half of the corer into the tree, he turns the handle one-half anti-clockwise and pulls out from inside a long, thin piece of wood].
>
> 'Here we have the "core"', Rob says triumphantly. [He inspects the core for three seconds]. 'This is a good sample because it has, approximately, one hundred rings.' [He puts the core inside a plastic straw and with a marker writes 'GLF', which stands for the name of the sampling site, 'Glen Falloch', and '1', the number.]
>
> 'It's your turn now. I will sample ten trees from here to the right, and you will sample another ten from here to the left. You will keep the numbers from 2 to

> 10. I will do from 11 to 21. We will meet at that fence over there in one hour. Agreed?'
>
> [There is a silence of three or four seconds. I feel daunted by the imprecision and openness of Rob's instructions. In front of me, there is an unquantifiable number of trees that look very much alike to me].
>
> I ask somewhat embarrassed, 'But how do I know which trees to sample?' Rob says, 'Just sample any tree … but be careful not to sample dead trees. You need to sample trees that look healthy and alive.' I insist and ask, 'And, what does a healthy tree look like?' Rob continues slightly impatient, 'Trees without scars, without resin and with a large green canopy. Off you go!'

One of the key aspects of the instruction detailed above is that my ability to learn how to sample trees adequately (including the skill of pre-empting future problems like measuring distorted rings from samples taken too low in the tree) was based upon my acceptance of Rob's authority as an expert teacher and the fact that I imitated his movements. A similar procedure occurred when I learnt to identify pieces of submerged wood in the lake. My entire body and, crucially, my sense of touch, were calibrated in accordance with Rob's as he was the fieldworker who was inside the water almost every day. When I asked him to describe his skill to me, he replied that 'you have to feel the wood', and prompted me to imitate his movements and identify the presence of submerged stumps with my feet. Once we identified a log, and managed to drag it out, it was crucial to identify whether the log was a Scots pine tree, and, if it was the right tree species, agree on the quality of the sample. There was huge expectation at this stage because dragging out a log could take us more than an hour. Rob, Björn and Miloš made such a decision by looking at the sample (see Figure 1.5d). They made comments such as 'this is a very sexy sample' (which meant that this was a good sample), 'this is worth 200 years' (which meant that this sample had 100 tree-rings) or 'this is shit birch, let's get rid of it' (which meant that it was a tree species that was not of interest). The selected samples were then labelled using masking tape (see Figure 1.5e).

The sharing of olfactory experiences also contributed to our training and to a sense of community among fieldworkers. On one occasion when I was sampling living trees with Miloš, I mentioned the amazing fragrance of fresh pine that came out of the tree. Miloš noticed my delight and said, 'Only people who spend time in pine forests and work with trees would recognise this smell.' Another example of the role of smell happened at the end of my first fieldwork day when we were all tired, looked pretty miserable and smelt quite badly. When we jumped into the car I noticed a distinctive stink of algae and sweat coming off mine and the others' bodies. I rolled down the windows to let in some fresh air, and Björn looked at me amused and said, laughing, 'You just need a bit more time to get used to the smell of doing fieldwork!'

Domestic rituals

Around 5pm, like in most office jobs, we finished our work in the forest and our attention shifted to preparing the evening meal at 'home' (the rented cottage where we slept, ate and socialised) and the associated rituals of domestic intimacy. One of the distinctive aspects of science done in the field, as noted by the historian Robert Kohler (2006: 67), is the combination of occupational and recreational activities. During the evening meal, the distinctions between 'home' and 'work', 'friends' and 'colleagues' became particularly blurry. The evening meal was the time of day when we returned to the rented cottage and shared a meal that one of us (either Björn or I, who gladly became the official chefs of the expedition) had prepared. Rob gave special priority to this socialising event. In fact, he told me that his main criteria in renting accommodation was to have a big dining table next to or inside the kitchen. The kitchen and the dining table were the main spaces where we ate, chatted, drank and played games.

The pivotal role of the evening meal was preceded by a series of activities. On our way to the cottage, we stopped at a supermarket to buy food for the evening meal and the next day's breakfast and lunch. We coordinated in the supermarket to find all the products we needed, and in an astonishing time of less than ten minutes, we met at the till where Rob was waiting to pay. When we arrived at the cottage, in turns, we took showers. As I was often the cook, I had priority for the shower. Those waiting for their turn gathered around the dining table, grabbed a drink and recapitulated the events of the day. The topic of conversation was often the number of samples we had produced (on a good day, we had between thirty and forty slices of wood from lakes and up to a hundred cores from living trees). The duration of the pre-dinner conversation depended on the time that the cook needed to prepare the meal. During this time, Rob and the rest of the team also cleaned and prepared the equipment for the following day and conversed about research.

The evening meal also developed according to a set of routines. The meal generally consisted of a high-calorie main course (pasta, rice or meat with vegetables) and a dessert with generous amounts of wine and beer. The cook received a public appreciation from the group at the beginning of the meal ('Thank you very much' or 'This is delicious!'), and a lively chat, steered by Rob who always sat at one end of the table, developed throughout the meal. The evening meal was a space where gossip circulated and fieldworkers shared opinions about colleagues. During or after the meal was also a time when we recalled the daily arduous experiences in the field and transformed them into adventurous stories at which we all laughed. One particular evening, I was the focus of collective laughing because Rob circulated a picture of me inside the lake where I looked visibly distressed ('It was freezing cold,' I said in my defence). After dinner, during the final hours of the fieldwork day, we often played cards and drank whisky. Before we all went to sleep around 11pm, Rob reminded us of the objectives and distribution of tasks for the following day. In this way, he made sure that everybody was ready to go back to work after a few hours of recreation around the dining table. We all went to sleep in gendered bedrooms. We had been together for twenty-four hours. Such was the intensity that fieldwork imprinted on me that every single night of the fieldwork expedition, I dreamt about the day's events in the field.

Conclusion

This chapter has detailed the fieldwork practices involved in producing samples for a dendroclimatological project. Rather than removing themselves from the process of deciding where to sample and what samples to use, generations of dendroclimatologists have sought to vindicate their professional judgement as an alternative strategy to 'random sampling' for achieving scientific objectivity (Daston and Galison, 2007: 309). It is a truism in dendroclimatology that 'purposive sampling', while not being free of biases, is the most appropriate strategy for generating good samples that show 'sensitive' tree-ring patterns and convey useful climatic information. This is why one dendroclimatologist told me that, 'There's no way around it' and Rob thought that no dendroclimatologist in his or her right mind 'would go to the tropics to study glaciers'. As shown in the two maps of the network of sites in the Scottish Highlands, the Scottish Pine Project was also based on, and in turn confirmed, the premise that the strategic choice of Scots pine trees, as previously demonstrated by Malcolm Hughes, was a valid and justified sampling strategy.

Precisely because dendroclimatologists uphold their professional judgement as a standard of scientificity and objective sampling, the production of credible samples in the Scottish Pine Project was simultaneous with the constitution of a cohesive community of expert fieldworkers. In dendroclimatology, pieces of wood become 'scientific' because they are produced by 'experts'. Being an 'expert' fieldworker is not an attribute that one acquires alone, but an ascribed status that other experts recognise and attribute while engaged in the actual doings in the field (see Coopmans and Button, 2014 for a discussion of how 'expertise' is constituted). As the leader of the Scottish Pine Project, Rob was responsible for ensuring that the fieldwork team learnt and exhibited the right skills and moral character necessary to be acknowledged as experts by himself and by the broader community of dendroclimatologists that recognised Rob as one of their own.

Fieldwork effectively served to guarantee the trustworthiness of the wood samples within Rob's team in that it generated a culture of trust, which confirmed and valorised Rob's personal and professional judgement – his decisions about which areas and trees to sample, which samples to retain and which to reject – and formed an important basis for group cohesion. Through participation in the rituals of labour and recreation in the field and at home, Rob was able to inculcate his skills, values and reasoning. More specifically, with the timetable and assignation of fieldwork tasks and budget, Rob conveyed his expectation that participants should take their fair share of hard work and bear this labour of division with goodwill if they wanted to be part of the moral economy regulating the exchange of tree-ring data. During our daily walks, Rob transmitted the importance of developing a deep personal knowledge and perceptual intimacy of the area and the surrounding trees (for a fictional and highly acclaimed account of the close relationships established between people and trees read Richard Power's (2018) *The Overstory*). In each hiking trip and tireless hour spent in the lake banks surrounded by midges, we learnt that fieldwork involved personal sacrifices. Through Rob's instructions, students like myself learnt to calibrate their senses (including one's reaction to the odour of sweat). Finally, through the rituals of preparation and sharing of the evening meal, participants created a shared memory of fieldwork.

Being and working *physically* in the field and at home was an opportunity for Rob to observe and confirm his students' and colleagues' acquisition of background knowledge about the ecological and historical features of the sampling sites and the growing conditions of sampled trees. In other words, Rob expected the members of his fieldwork team to generate 'metadata'. While metadata are commonly regarded as static and well-codified products, in routine scientific practice like fieldwork metadata are manifested as ephemeral, informal and incomplete objects (see Edwards et al., 2011: 684). This is why, when I asked Rob about the importance he attributed to fieldwork, he established a connection between fieldwork and the tree-ring data generated at later stages ('We've been in the sites, we've done the data, and we know them so well'). This is also why Rob insisted that his undergraduate students participated in fieldwork as he thought 'it is important that they see for themselves where the data they'll generate come from.' The fact that Rob conditioned the sharing of tree-ring data on participating in fieldwork ('basically, I share the samples with those who deserve them') is also indicative of the morality he associated with the development of first-hand experience of the local context of production of samples.

The creation of a group of fieldworkers who valued their empirically grounded and fieldwork-based expertise was crucial to the efforts of creating Figure 0.1, as will be shown in the following chapters. At later stages of their epistemic work, Rob and Miloš would refer to the knowledge or 'metadata' developed in the field site to interpret tree-ring data in a process akin to what Bruno Latour (1999: 24) calls 'circulating reference' (the word 'reference' comes from the Latin 'referre', which means 'to bring back'). For instance, when Rob and Miloš had difficulties crossdating some samples from the Alladale site, they initially relied on the knowledge of the logging history of the Scottish Highlands passed on by their field relations (estate owners and government officials) to interpret such an anomaly (they later concluded that logging had nothing to do with it). Similarly, Miloš justified the adjustments he made to a tree-ring model by appealing to his field experience of the wet conditions in the West of Scotland. In the controversy chapter, Rob refused the hypothesis formulated by a colleague as 'unrealistic' on the grounds that this colleague likely did not have sampling experience. Ultimately, all these examples show that the 'lab–field border' – as Robert Kohler (2002) refers to the cultural differences between indoor or laboratory science and outdoor or fieldwork science – is particularly fluid in dendroclimatology, as the background knowledge or 'metadata' generated during fieldwork cuts across the laboratory work done at later stages.

The strategy of invoking the experience gained in the field site as a means to justify subsequent interpretations of data seems to be characteristic of fieldwork-based sciences more generally (Kohler and Vetter, 2016; Kuklick and Kohler, 1996). In their study of petroleum geophysicists, Petter Almklov and Vidar Hepsø (2011: 552) argue that these scientists used their field trips to offshore reservoirs to create 'analogues' that informed their interpretations of remote data sources later in the office. The authors ask the very pertinent question, 'What comes back?' when these geologists have left the field site, to which Almklov and Hepsø give an answer that could also apply to dendroclimatology: 'The data are the objects of attention, both on field trips and in the office, but the answer to the question of what comes back from the logging trips is not primarily a set of immutable mobiles (Latour, 1987), but instead a group of professionals with an increased understanding of the context from which geological data are extracted.' In other words, what comes back after fieldwork is a community of experts. Crucially, individual geophysicists and dendroclimatologists are able to mobilise their field experience successfully because their respective communities accept such experience as a sound basis for establishing extrapolations.

The existence of an exclusive community of peers that willingly gave assent to the quality of Rob's scientific judgement and granted him exclusivity over the relations, sampling sites and data produced bears some relevance to public debates about openness and secrecy in climate science. After the hacking of the climate science emails in November 2009, many inquiry reports, individual scientists and scientific institutions lamented that the stolen emails had shown failures to abide by Freedom of Information legislation that requires all publicly funded institutions to provide access to their data and results, and defended openness as the basis for science's capacity for self-correction and peer-review (Ramírez-i-Ollé, 2015a: 398). The then president of the US National Academy of Sciences insisted that 'Clarity and transparency must be reinforced to build and maintain trust – internal and external – in science' (Cicerone, 2010: 221).

Scientists and their representative institutions commonly present openness and transparency, no matter how qualified, as ideal scientific norms, and secrecy and intimacy as unjustified and deviant. Yet, given the meticulous work and personal investment and sacrifice required to produce samples like those of the Scottish Pine Project, how can we be surprised that some scientists develop a sense of attachment and moral ownership towards the objects, places and people involved in their production? If one understands that restricting access to data or allowing temporary monopoly rights over the exploitation of samples are the means by which scientists like Rob protect junior researchers and incentivise good team work, why do we, as a society, still expect that scientists will share their precious goods with strangers and abide by Freedom of Information legislation? How can it be easy to trust people who have not developed an embodied and intimate knowledge of the sampling areas or 'metadata' to make appropriate use of the resulting data? To be clear, I am not making a case against transparency laws (even though I think there are good reasons for opposing auditing measures that generate wasteful games of compliance and undermine public trust; Power, 2005: 10; O'Neill, 2002). Instead, I hope the reader will conclude from this chapter that it is not surprising that climate scientists might actively oppose or find it difficult to comply with transparency laws, as secrecy and exclusivity have a prominent, and sometimes functional, role in their work.

While science is commonly seen as being about 'unveiling' secrets about nature and doing so in the open, the reality is that, for better or for worse, reliable scientific knowledge is mundanely done in private. Large parts of science are kept behind closed doors for military or industrial reasons, because of everyday competition between scientists or, as shown in this chapter, because temporary exclusivity is part of an internal reward system that guarantees quality work within a scientific group. The more normalised, protective practice of 'peer-review' is also paradigmatic of the existence of a culture of secrecy in contemporary science (as Røstvik and Fyfe, 2018 have shown in their historical analysis of the journals of the Royal Society, women have been particularly affected by such form of exclusivity). More generally, concealment and openness have co-existed throughout the history of science and within the individual careers of prominent scientists. For instance, Isaac Newton kept his alchemical work, which he valued greatly, secret, and only made it available to an elite (Golinski, 1988). Likewise, Galileo sent his telescopes to cardinals and princesses who supported his work, but refused to show them to competitors like Kepler (Biaglioli, 2006). More generally, the sociologist Georg Simmel (1906: 441) explained long ago the inevitable existence of 'secret societies', as 'in all relationships of a personally differentiated sort there develop, as we may affirm with obvious reservations, intensity and shading in the degree in which each unit reveals himself [sic] to the other through word and deed'.

Therefore, I hope to have demonstrated that the widespread idea that secrecy and privacy are pathological practices and inherently bad for the conduct of climate science should be qualified. Openness and secrecy are not absolute goals, but instead, as Brian Balmer suggests (2012: 17), we should think of them as context-specific, as 'degrees of openness and secrecy, choices over what to reveal or conceal, point to a geographical analogy, that openness exists as a zone between two others: the initial secret, then what is revealed about that secret, but also what is held back and remains out of sight'. The fact that secrecy is not essentially bad does not mean that it should be applauded. Given the emergence of a 'culture of transparency' in many Western countries (Schudson, 2015), it is likely that demands for scientific openness will keep growing. The challenge for policy-makers is, as Sheila Jasanoff (2006: 22) foresees, to develop legislation that 'maintains a desired balance between openness and secrecy'. To achieve this goal, legislators should attend to epistemographic evidence, such as that provided in this chapter, that foregrounds the practices and reasons that constitute the secretive work that they intend to open up to public scrutiny.

2

Dendrochronology

After finishing fieldwork – while driving back to Edinburgh where Rob, Miloš and I lived and where Björn and Hans caught their planes to return to Sweden – we always stopped at the University of St Andrews to leave the samples we had painstakingly produced the week before. Since Rob had accepted the position of Lecturer at St Andrews University in 2007, he had been negotiating to have an exclusive room to store and work with the samples. Much to his regret at the time, all he had achieved was a bigger office room that he also used as a storage room for the samples (Figure 2.1). 'Here, we have the entire Scottish Highlands', Rob announced triumphantly, pointing to the drawers full of cores and the piles of slices of subfossil wood.

From the moment we left the samples in Rob's office, Miloš became responsible for creating data from them in a procedure known as 'dendrochronology'. From each tree sample, dendrochronologists generate a series of tree-ring measurements and then combine several of these series to produce what is known as a 'tree-ring chronology'. Essentially, dendrochronologists aim to identify the precise year of formation of each measured tree-ring in order to produce carefully dated tree-ring chronologies. The sequence of dated tree-rings results from the fact that trees *generally* grow one layer of wood every year (as I explain below, it is possible that, in some years, trees produce no ring at all or that they produce more than one ring).

Overlapping successively older tree-ring measurement series into an averaged 'master chronology' – a procedure known as 'crossdating' – is considered the foundational practice of dendrochronology (see Rob's PowerPoint slide, Figure 2.2). As I describe in Chapter 4, creating long tree-ring chronologies is crucial for dendroclimatologists because the longer the chronology, the longer the climate reconstruction. Precisely, Rob and Miloš's goal at this stage was to create a master tree-ring chronology from all the different types of sample across the Scottish Highlands (from lakes, historical buildings and living trees) that extended further back in time than the climate reconstruction for Scotland created by Hughes in the 1980s (Hughes et al., 1984), which reached the eighteenth century.

Textbook authors employ various metaphors to explain how dendrochronologists use tree-rings as a source of climatic data (dendroclimatology is only one of the possible applications of dendrochronology; hence, while all dendroclimatologists

2.1 Rob's office stored the wood samples from which Miloš generated data in the form of series of dated tree-rings or 'tree-ring chronologies'.

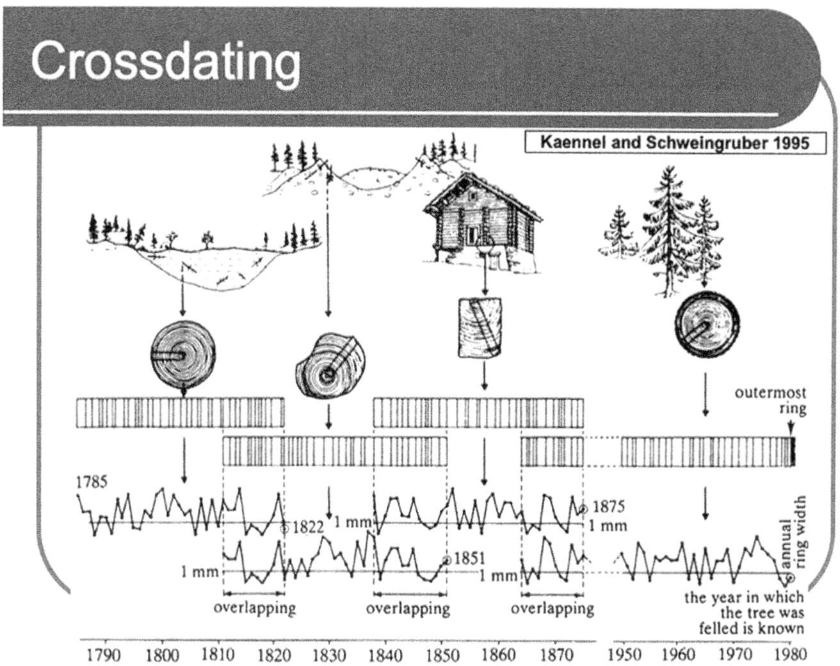

2.2 Dendrochronology consists in creating accurately dated and longer tree-ring chronologies by cross-matching ring patterns and dated tree-rings across different samples.

are also dendrochronologists, not all dendrochronologists are dendroclimatologists; see Chapter 3). Tree-ring measurement series are compared to a 'bar code with varying widths of lines representing each year' (Speer, 2010: 12). Andrew Douglass – described as 'the undisputed father of dendrochronology' (Speer, 2010: 37) – compared the pattern of narrow and wide tree-rings to the Morse telegraph code of dots and dashes. Harold Fritts (1976: 19) followed up this metaphor: 'In much the same way, the sequence of narrow (dots) and wide (dashes) rings in a sensitive ring series conveys messages about the life of the tree.'

Rob, Miloš and the dendrochronologists I talked to about this matter were adamant that doing dendrochronology did *not* involve counting tree-rings. The website of one of the oldest dendrochronology laboratories –founded by Andrew Douglass in Tucson, Arizona – precisely introduces dendrochronology by asking the question, 'Why not just count the rings?' I asked a similar question to Rob when I was training as a laboratory technician: 'Could we not create tree-ring chronologies by counting the total number of rings in each sample, and then assigning the calendar year to each tree-ring by counting backwards from the outermost ring, laid down in the year when the tree was sampled?' Rob categorically responded to me, 'Dendrochronologists don't count; foresters count.' Rob was not unique in characterising foresters' ring-counting practices as improper and unscientific. In a lecture I attended in Tasmania (January 2014), a local forester gave a presentation about a few chronologies he had created with local timber. When I asked one of the attendees, a dendrochronologist, his opinion about the talk, he said, 'It was interesting because he knows these forests very well, but clearly, he is not doing the same thing that we do. I don't think he did any crossdating.'

Dendrochronologists insist that ring-counting is not a valid procedure for creating accurate tree-ring chronologies because it cannot account for the potential existence of tree-growth anomalies referred to as 'false' and 'missing' rings. In dendrochronology textbooks, false and missing rings are given a physiological and ecological explanation. As described by Fritz Hans Schweingruber (1988: 47), 'trees and shrubs in temperate and boreal regions generally form one ring a year. The occurrence of clearly marked seasons in these regions means that very little variation in growth occurs within any given year. There can, by contrast, be enormous variations in growth in trees and shrubs in arid and semi-arid regions. This is the result of the unevenly distributed precipitation. A ring may be formed every year, or there may be no growth whatsoever [missing ring], or the tree may even appear to form two or three rings in a year [false rings]. For samples in these regions age can be determined only by cross-dating.'

The 'epistemological conundrum' that Miloš and Rob, and all dendrochronologists, face at this stage of the production of dendroclimatological knowledge is to produce accurately dated tree-ring data, given all the uncertainties related to the potential presence of missing and false tree-rings. An added challenge specific to the Scottish Pine Project was that, as seen in the previous chapter, Rob's team worked with samples from lakes and buildings. Determining the exact year in which submerged trees or archaeological beams ceased growing and the year in which the last ring was laid down – the starting point for dendrochronology – is particularly difficult, if not impossible. In this chapter, I shall describe how Miloš, Rob and myself as a voluntary technician resolved these challenges and eventually created an 800-year-long tree-ring chronology from Scotland by reading and representing tree-rings and synchronising tree-ring data.

Reading tree-rings

The first time Miloš asked me if I were able to *see* a specific tree-ring that he was showing to me, I did not know what to look at. I was still in shock after opening the plastic bag with the samples from fieldwork: I did not expect the strong smell of rotten eggs coming out from the samples after being bagged for weeks. I was also in shock because, when looking at the code attached to each sample, I had flashback memories of fieldwork. I could remember the exact time and place of origin of almost every sample. In my first day as a voluntary technician in the laboratory, Miloš and I spent considerable time remembering anecdotes from fieldwork.

It quickly became clear to me that *seeing* tree-rings was not a simple task; I had to develop the 'professional vision' of the dendrochronologist (Goodwin, 1994) or a 'dendrochronologist's eye', which essentially involved learning to *interpret* and *read* the dark bands that I perceived on the wood. The process of interpretation or reading of tree-rings is double-sided. On the one hand, it is 'theoretically–culturally laden' by the concepts (e.g. a 'tree-ring') and laboratory practices of generations of dendrochronologists that enable the perception of the wood. On the other hand, it is 'empirically–materially laden' by the specific tree-ring structure under study. As the sociologists of science Barry Barnes, David Bloor and John Henry (1996: 28) explain, 'The mind of the individual is the point of contact between our physical environment and social environment. Interpretation is where nature and culture come together.'

Once samples arrived at the laboratory, our main goal was to enhance the visibility of tree-rings. As the sociologist of science Michael Lynch (1985) has shown, when scientists bring natural specimens into the laboratory for study, they have to transform them into 'docile objects'. In the case of the dendroclimatological work conducted in Scotland, this 'domestication' started at the field site when fieldworkers tagged samples and cut wood pieces of a certain size in order to fit them in Rob's office. When these samples arrived in the laboratory, we sought to tame them even more. In order to remove wood resins and to measure rings from wooden cores, we submerged the cores in jars of acetone and glued them onto prefabricated wooden mounts (Figure 2.3a). Sometimes we had to break the core into small pieces so that the tree-rings would look 'straight' and have the correct orientation on the wooden mounts (Figure 2.3b). As Miloš explained to me, tree-rings could come out 'twisted' from using poorly sharpened corers. With regards to subfossil samples, we tried to stop the deterioration of wood and the appearance of mould by air-drying the slices of wood and using a small blade to remove part of the rotten surface, and adding chalk to further enhance the contrast and visibility of rings (Figure 2.3c). Throughout this process it was very important,

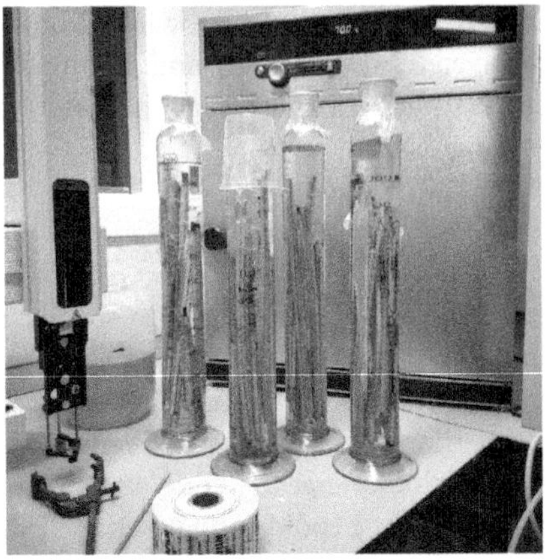

2.3a

2.3b

2.3c

2.3 Being able to read tree-rings involved a series of laboratory practices that 'domesticated' samples and made tree-rings more visible and amenable to interpretation.

Miloš insisted, to keep track of the identity of samples and to assign the correct sample code by attaching each core to a numbered metal wire and writing the code on the wooden mounts. In this way, we would not lose the connection of the samples with the place they would be taken to represent at later stages (Latour, 1999).

After preparing samples for visual examination with the microscope or with the computer, reading rings still involved many uncertainties with regards to the exact boundaries of a ring. On an ordinary day in the laboratory, one of the most frequent questions I asked Miloš was: 'Is *this* a ring?', while pointing to the ring in question. If I was measuring rings in subfossil wood with a microscope, I used a pointer to mark the exact ring that I was doubtful of. Neophytes were asked to count tree-rings in groups of 10 following an established code (one dot every 10 rings, two dots every 50 rings and three dots every 100 rings). I found it much easier to discuss my doubts with Miloš in front of the computer screen rather than in front of the microscope because we could use our hands and body to indicate the boundaries of the ring on the digitised image of the sample (see Vertesi, 2012, for an account of the importance of bodily communication in science). In this way, I learnt to read tree-rings through verbal and bodily interactions like the following:

> Meritxell: Is this [pointing to the computer screen] a ring?
> Miloš: Mm, let me see
> Meritxell: [I stand up to let Miloš sit in front of the screen]
> Miloš: [sits down] (pause of 3 seconds) [slight turn of the head] [zoom in and out of the picture]
> Miloš: This is very tight [he leans towards the screen]
> Meritxell: Yes, I know
> (pause of 6 seconds)
> Miloš: I would say this [pointing to the screen] is a ring
> Meritxell: Really?
> Miloš: Yes, you can see the dark band of the ring here [moving the finger in circles in front of the screen]
> Meritxell: Okay

Another way in which I learned how to read rings was through the use of diagrams and exemplary images in Rob's lectures. He typically used images that highlighted the distinct boundaries of a ring (Figure 2.4). Rob employed these images to re-create in the minds of students the act of identification of tree-rings that took place in the laboratory – like the conversation between myself and Miloš described above – which undergraduate students could not directly witness in the classroom (the sociologist and historian of science Steven Shapin, 1984 refers to this process as 'virtual witnessing'). All the verbal, bodily and written demonstrations and instructions that we, as students, accepted from authoritative teachers such as Miloš and Rob organised our perception of the ring patterns on the wood and constituted the professional and competent vision of future dendrochronologists who could see and classify a dark band as a 'tree-ring' or a 'non-tree-ring'.

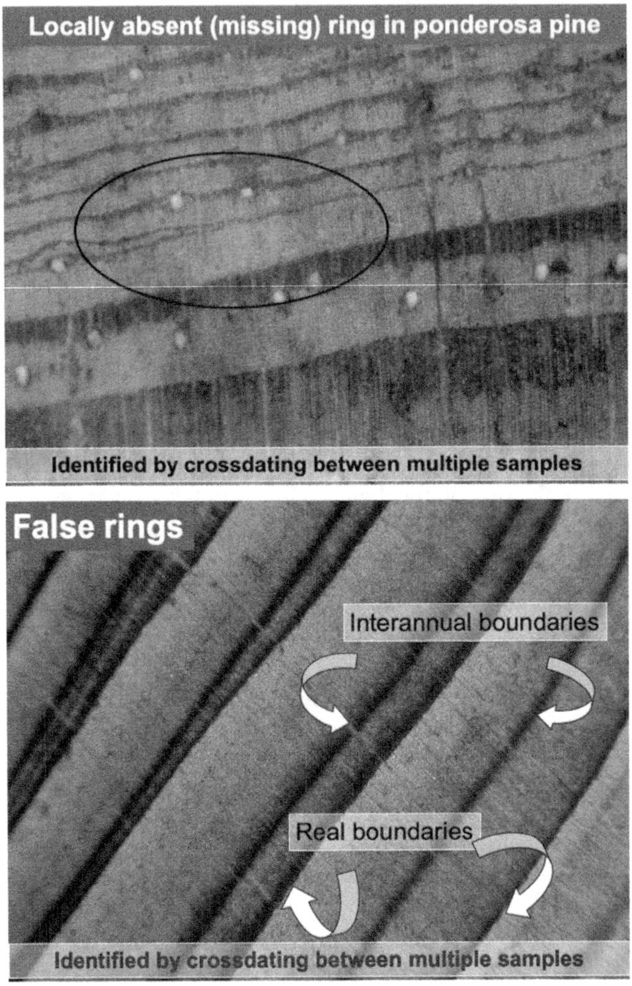

2.4 Rob used this picture in class to re-enact in the students' minds the bodily gestures and conversations taking place in the laboratory that enabled the interpretation of tree-rings.

Representing tree-rings

Once I learnt to read tree-rings, Miloš also taught me to generate accurate measurements of two physical properties or 'parameters' of a tree-ring: its width and reflectance. Traditionally, tree-ring chronologies have been created by measuring the width of annual rings. The emergence of newer technologies and the geographical expansion of dendroclimatology around the world have meant that additional information from tree-rings is now also measured. Newer tree-ring parameters include the measurement of ring density and the reflectance or brightness of wood. As Rob could not afford the workforce or the technology to generate density data (instead, he relied on his Swedish colleague Björn to produce density), he refined a cheaper parameter or proxy called 'Blue Intensity' with the help of colleagues around the world (Ramírez-i-Ollé, 2018).

Ring-width and blue intensity measurements 'represented' – in the sense of acting as substitutes of – the wooden samples at later stages of the work. As I shall explain in the next section, Miloš would only go back to the physical samples (or their scanned images) if he struggled to crossdate. Scanned images had the advantage of being easier to store than samples. In fact, keeping physical samples was often a problem for dendrochronologists. While Rob did not yet have storage problems (as he still had space in his office), he travelled to Canada in March 2015 to help his doctoral supervisor with this problem. Rob told me that despite the fact that his supervisor's samples are among the oldest, and hence most valuable, in the world, his supervisor had not been able to convince the university officials to keep the samples after he retired. As a result, Rob 'inherited' his supervisor's samples and travelled to Canada to collect them and archive them in a dendrochronology laboratory in New York (the 'Lamont–Doherty Earth Observatory', commonly called 'Lamont') where Rob had worked as post-doctoral researcher and maintained an affiliation.

The key aspect to consider when generating accurate measurements or representations of tree-rings was to not overrule the experienced human eye. Miloš and I used two machines for generating measurements (Figure 2.5). First, we employed a measuring stage that worked in conjunction with a microscope, cross-hairs and a digitising recording device. As we moved the stage with a little handle and the cross-hair coincided perpendicularly with the ring boundary, we pressed a button and a device recorded the distance travelled by the measuring stage in millimetres, which was later recorded by a computer. Secondly, we employed computer software that detected the ring boundaries and placed measurement points automatically on the scanned image. With a mouse, we identified one ring boundary, and with the use of coordinates the programme detected and measured the subsequent rings automatically.

Miloš insisted that I should take into consideration different criteria for generating tree-ring measurements. If I were generating ring-width data, I had to measure the ring at the point where it looked more 'proportional'; that is, the area where the width of the ring looked equally wide or narrow (as trees do not grow uniformly and tree-rings are not perfect concentric circles). If I generated blue intensity data, I had to be careful in delineating the darkest parts of the rings and avoid measuring sections that appeared anomalously bright or dark. Using the words of the renowned dendrochronologist Michael Baillie – author of one of the longest tree-ring chronologies (Box 3) – one could say that the tree-ring measurements we produced with Miloš carried out our 'intellectual fingerprint' insofar as our goals (the measurement of specific parameters) and experience (as a student or as an expert) made a mark on the resulting data. As Bailie explains it:

> When a dendrochronologist measures the widths of growth rings in a sample, he or she has to make multiple decisions with respect to the starts and ends of the rings, problem rings, and so on. Repeated measurement of the same sample,

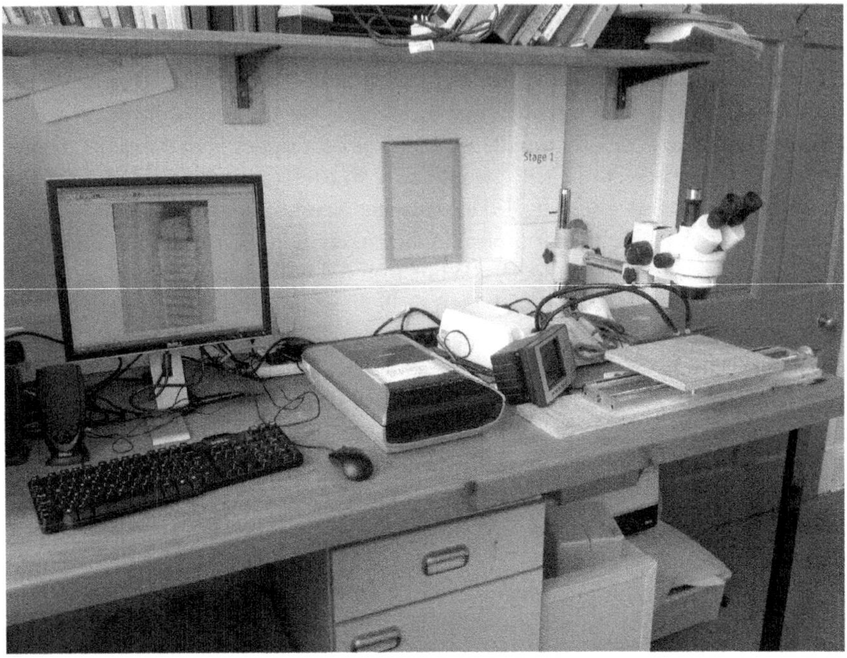

2.5 The measuring machines used for generating tree-ring width and blue intensity data, which represented and substituted the physical samples at later stages.

> will not give exactly the same measurements. The number of rings must be the same, but the actual measured widths will not be. This means that the ring pattern of a tree-ring sample carries the 'intellectual fingerprint' of the dendrochronologist who measured it, every bit as much as this text carries my intellectual fingerprint. (Baillie, 2010).

Establishing the accuracy of tree-ring measurements was not about upholding an absolute standard. Miloš never expected my measurements to be exactly the same as his, but if he deemed the difference to be too large (on whatever grounds), he asked me to measure the tree-rings again. On one occasion, Miloš made explicit this criterion of 'relative accuracy', when he congratulated me by saying, 'Well done! This almost looks like I had done it myself!' Assessing the quality of measurements in relation to the measurements of experts seems to be a common practice in dendrochronology. Harold Fritts (1976: 251) explains in his textbook that he devised a 'test of measurement accuracy', which can be made 'by comparing measurements of particular operators to those of experts'. Rob told me of the existence of a more recent software ('Verify5') that was also designed to do such tests, but he had never used it because, as he explained, 'one must have some faith in students' (however, Rob often used the measurement software package used by his students, 'CooRecorder', to check their measurements). Neither Rob nor Miloš used an 'accuracy league' displayed on the wall to compare the students' measurements, as some laboratories seemed to do in the 1990s (Pilcher, 1990: 45).

Box 3

How much knowledge do I take for granted?

In March 2015, I published an article in a new and relatively unknown open-access journal where I made a claim about Michael Baillie's work that, to my surprise, became slightly controversial. More specifically, I said that 'the longest tree-ring chronology created anywhere in the world was built by European researchers using Irish oak trees, currently reaching back about 10,000 years' (Ramírez-i-Ollé, 2015b). I made this claim using Baillie's published work (1995), quoted in all dendrochronology textbooks. As usual, prior to publication, Rob and Miloš had read my article but did not amend my claim about Baillie's chronology. All dendrochronologists seemed to accept that Baillie's chronology was the longest in the world, and so did I without being aware of it.

A few days after my article was published, Rob included me in an email conversation with Douglas Keenan – who describes himself in his website as 'an independent mathematical scientist'– who questioned the accuracy of Baillie's chronology. Keenan's email included a detailed examination of the potential dating problems in Baillie's chronology and finished by asking Rob, 'How confident are you that the statement is true?' in reference to my claim about Baillie's chronology being the longest in the world. Rob replied by summarising some uncertainties of Baillie's chronology and responding amicably to 'Doug': 'So – to answer your question – I don't honestly know, but as I am sure Meri [me] would be happy to hear – I have faith in my colleagues to get it right. If there is disagreement, then both parties should work together and try and resolve the issue.' I followed up this conversation by sending an email to both 'Doug and Rob' and thanking them because 'I've learnt a few more things about tree-ring dating in these exchanges.'

This incident made me aware of an existing, if marginal, criticism of a dendroclimatological fact and forced me to consider how, if at all, I should foreground it in my work. How can I write an epistemography of a scientific object without falsely making it appear more factual and credible given the existence of disagreements? The way I tried to resolve this problem is by adopting a well-known epistemographic claim. Bruno Latour and Steve Woolgar ([1979] 1986: 75) claimed that scientific objects go through various stages of facticity, linguistically expressed in the way a statement is gradually transformed from an issue of hotly contested discussion into a well-known, unconventional remark freed from the circumstances of production. While some statements comprise conjectures ('It should not be forgotten that…'; 'These results are temporary'), others are unequivocal and factual ('X and X have reported that…'; 'It is largely known that…'; 'A has a relationship with B'). Latour and Woolgar studied the process of fact-making over time, but their argument, I suggest, also applies to extant claims. In this way, my claim about Baillie's work could be situated in a continuum that progressively includes more contextual information:

> *The longest tree-ring chronology reaches back about 10,000 years.* [This is the factual language that dendrochronologists would use, which omits contextual information such as the subject, the sources and the 'created' nature of the chronology. I obviously avoided this formulation.]
>
> *The longest tree-ring chronology created anywhere in the world was built by European researchers using Irish oak trees, currently reaching back about 10,000 years.* [This is the actual sentence that I included in my article. It includes references to human agency ('created' and 'built') and it accurately attributes the production of this chronology to a collective ('European researchers') rather than to Michael Baillie alone.]
>
> *The longest tree-ring chronology created anywhere in the world was built by European researchers using Irish oak trees, currently reaching back about 10,000 years (there is some debate about its exact length; Keenan and Wilson 2015, personal communication; Baillie 1995).* [This formulation is better than the previous one because it shows the ongoing making of a dendrochronological fact or epistemic object. It includes a reference to Baillie's work so that the reader can check the original source where some of the uncertainties discussed by Rob and Doug appear].

If epistemographers are aware of existing challenges to the epistemic object and practices under study – no matter how irrelevant these challenges might seem to our subjects – we should document them. Our goal is to offer an accurate analysis of knowledge-making, and this includes, as Latour and Woolgar ([1979] 1986: 80) argue, showing that 'changes in the type of statement provide the *possibility* of changes in the fact-like status of statements'. In this book I have sought to indicate the sources of potential changes in Figure 0.1 by presenting 'epistemic conundrums' at the start of each chapter. Some of these conundrums and limitations were identified externally – by outsiders to the Scottish Pine Project or to the community of dendroclimatologists more broadly – whereas other conundrums originated internally. As I illustrate in the conclusion section of Chapter 5, the interplay of influences coming from the 'inside' and the 'outside' of the scientific community in question explains the dynamism of epistemic objects and should be a primary focus of epistemographic analysis (Shapin, 1992).

Synchronising tree-ring data

When I finished measuring my first sample I was exultant because I (wrongly) assumed that I had created my first chronology. My confusion stemmed from the fact that the computer automatically attributed calendar years to each measured tree-ring. For instance, if I had measured fifty-five rings in a core from a tree sampled in 2015, the machine assigned a year (2014, 2013, 2012 …) to each measured ring counting backwards from the bark. When I finished measuring, I proudly announced to Miloš, 'I'm done with my first chronology!', to which he replied ironically, 'No, you're not! You can't just accept the dates that the machine assigns to tree-ring measurements. We've got to crossdate.'

Miloš crossdated tree-ring data with the help of the computer and 'correlation coefficients' that measured the similarity between ring-width or blue intensity measurements from different samples. Miloš used both a Microsoft Disk Operating System (MS-DOS) software commonly used in dendrochronology called 'COFECHA' and a more recent program called 'CDendro'. The difference between the two packages was, essentially, that CDendro could graphically represent series of tree-ring measurements and correlation coefficients (Figure 2.6), which, as Miloš said, 'makes it easier to compare the data'. Miloš used these two packages from his home and, in this way, his place of residence also became his (and my) place of scientific research in addition to the laboratory in St Andrews (for the importance of the household in early seventeenth-century science in England see Werret, 2019; Shapin, 1988). Once per week, for a year, I went to Miloš's house, sat next to him, observed and took notes.

The iterative cross-checking of calendar dates and tree-ring measurements is a process that dendrochronologists call 'chronology-building', as they progressively create an averaged master chronology by building up a dataset of similarly correlated measured samples. With samples from living trees, Miloš started by comparing measurements of pairs of replicated samples from the same tree (1A against 1B; 2A against 2B, etc.) and then proceeded by comparing measurements of different trees from the same site. With all these comparisons, Miloš would 'get a feel' for the correlation coefficient he could expect in a sampling site; what he often found was that correlations between measurements from the same tree were higher and better than those from different trees. Miloš then used this benchmark to distribute individual datasets into 'good' and 'bad' computer folders. The 'good' folder included individual measurement series that Miloš thought were sufficiently similar to the master chronology-in-the-making, whereas the 'bad' folder included poorly correlated measurement series. As I detail below, crossdating subfossil samples was far more challenging, and Miloš and Rob created a system for validating such chronologies.

While Miloš was open to the possibility of not finding any cross-matching between tree-ring measurements, his work was driven by the expectation that rings from trees grown in similar environments should crossdate, and in case they did not, tree-ring measurements would have to be synchronised. Miloš actively looked for 'mismatches', which he identified, both visually and statistically, as weakly correlated segments in CDendro and COFECHA (see vertical line in Figure 2.6). When this happened, Miloš went back to the digitised image of the sample and identified those sections of rings where the statistical correlations 'broke down' or the mismatching occurred. He then started looking for explanations for this asynchrony: 'Perhaps I missed a ring or the measurements of the rings are not perfect.' Only after visually re-examining the sample or its scanned image, or

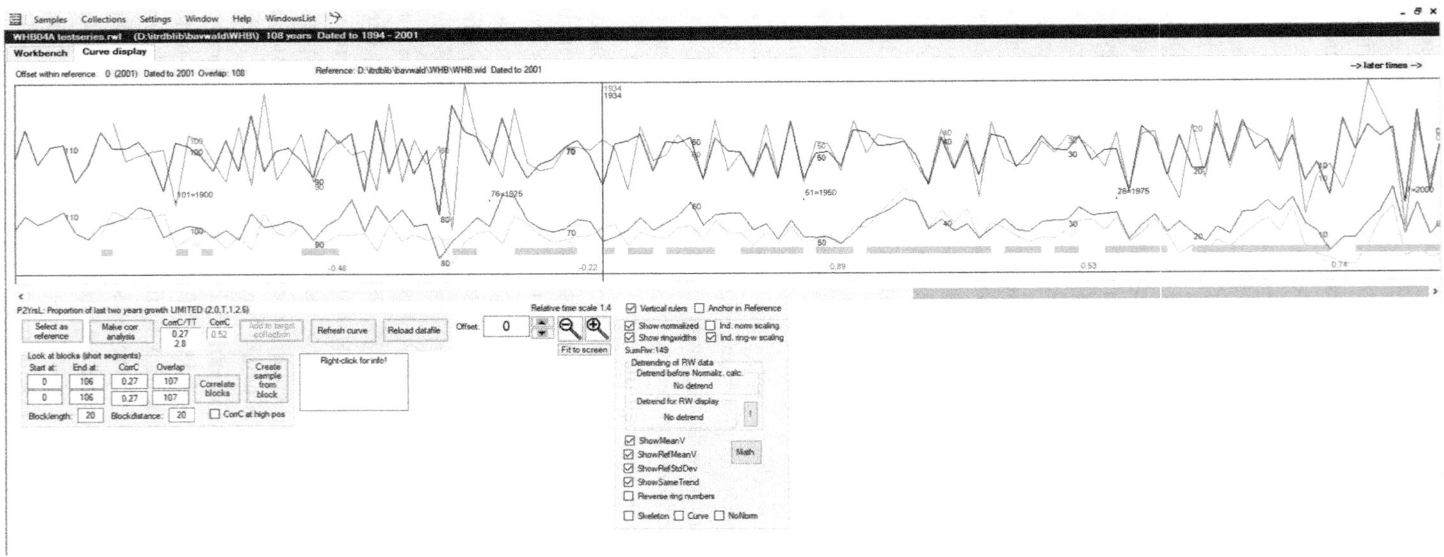

2.6 Crossdating tree-ring chronologies involved inferring and interpreting the presence of false rings or missing rings and then synchronising measurements across datasets.

sometimes even re-measuring and noting no improvement in the correlations, did Miloš consider the possibility of the existence of false or missing rings: 'If you see a compressed section of rings or a section of the scanned image with very narrow or diffuse rings, it is plausible to consider that there could be an extra missing ring that we can't see.' Miloš then right-clicked twice with the mouse at the section of one of the graph lines where there was a low correlation, and in this way he added two extra measurements or 'missing' tree-rings. Miloš then pointed to a little box on the screen that showed that the correlations were now significantly higher and the measured series lines matched.

Miloš always insisted that 'You cannot allow statistics to manipulate you.' What he meant is that one cannot accept the 'offset years' recommended by the software as the real number of missing or false rings, which, if included or subtracted, increased the synchrony among measured series of tree-rings and the correlation coefficients. Miloš told me that he was not content to accept the default correlation coefficient threshold arbitrarily established by the software programs (in COFECHA the threshold was 0.33 and in CDendro it was 0.4) as evidence of accurate tree-ring dating. 'You cannot just go with what the program tells you to include or to remove', Miloš explained. 'You need to understand how reasonable the recommendation made by the computer programs is.' In appealing to the need to develop an informed opinion about statistical matches, Miloš was in line with what the acknowledged founder of dendrochronology, Andrew Douglass, argued dendrochronologists should do: 'There is no mechanical process, no rule of thumb, no formula, no correlation coefficient, to take the place of this personal comparison between different ring records; the operator does not dare to seek relief from his responsibility' (cited in Pilcher, 1990: 43).

If Miloš found samples that required 'too many tweaks' or too many adjustments for false and missing rings that he deemed unjustifiable, he considered these samples 'uncrossdatable'. This was the case with 50 per cent of the samples from the 'Alladale' site in the North-West of Scotland; in this exceptional case, Miloš found that the pairs from the same trees (1A–1B) did not even crossdate (as I explain in Chapter 4, Miloš and Rob initially interpreted the anomalies in the Alladale site in relation to a broader anomaly they called 'disturbance'). With subfossil samples, Miloš warned me of the higher possibility that the 'offset years' and statistical matches offered by the computer were 'spurious' and did not reflect a real crossdating. Miloš used the metaphor of a 'jigsaw' to explain the difficulties in creating tree-ring chronologies from subfossil wood:

> Imagine that you've a jigsaw of 10,000 pieces, but the picture you're creating only involves some of those pieces, which are the subfossil samples that we have from Loch Gamhna. You might only be able to create a little part of the jigsaw out of the thousands of pieces. But you don't know which pieces you have; you don't know how they fit together or if they actually fit together because the subfossil samples do not necessarily come from the same forest – some of the logs could have originated in other areas and ended up in the lake. You don't even know what the total number of jigsaw pieces is. In fact, every year, this number gets bigger and bigger as we collect more samples.

Miloš partly resolved the uncertainties regarding the dating of sub-fossil samples by relying on an established technique called 'radiocarbon-dating', which offers a probable age range of the wood based on the analysis of the concentration and decay of radioactive carbon isotopes (^{14}C). The calculation of these date estimates requires expensive machinery and specialised skills that Rob and Miloš did not possess. Therefore, Rob occasionally sent a selection of samples to an external laboratory and paid a considerable amount of money (£300 per sample) to obtain a radiocarbon date. Prompted by my questioning (Box 4), Miloš told me that he had no concerns about the method itself and the professionalism of the experts involved in calculating the radiocarbon dates. Even though Miloš was aware that the radiocarbon dates could be inaccurate in some cases, he was certain that 'the real radiocarbon date would actually be within the 95 per cent confidence range', and therefore he used these radiocarbon-dating estimates as the basis for crossdating subfossil wood and cross-checking his chronologies with Rob.

Box 4

How 'breaching experiments' turned against me

Throughout my research I employed a research method inspired by Harold Garfinkel's (1967: 35) 'breaching experiments'. Essentially, Garfinkel asked his students to document the 'breaching' or disruptive effects of social experiments that consisted of systematically distrusting people. For instance, one student questioned whether a bus driver would follow the assigned route and the bus driver became, quite understandably, upset with her. A married student, meanwhile, doubted her husband's account of why he had arrived home late the night before, which generated resentment between the couple long after she admitted the experiment. Garfinkel designed these experiments to show students the trust-dependency of interactions and the risks associated with distrusting the shared assumptions sustaining everyday life. Similarly, my intention in adopting Garfinkel's approach was to foreground the role of trust relations and background expectations in Rob and Miloš's work by temporally questioning them.

At the start of my research, I was often wrong in predicting the breaching effects of my interventions because I had not yet developed a good understanding of Rob and Miloš's culture and what they took for granted. On one occasion, after two months of working as a laboratory technician, I asked Miloš a question that I did not think would be controversial, but it turned out to be so. More specifically, I asked him whether he trusted carbon estimates; he looked at me, very surprised, and replied, 'Of course!' He continued, 'The methodology involved is pretty robust; it's a whole area of science. We don't do it ourselves. It's not part of our job. We just need to know what it is and how to interpret it.' As I was listening to Miloš's response, I realised that his reliance on the carbon-dating method was a perfect example of the routine trust that I sought to make visible with my breaching questions. As I initially did not want to reveal my experimental strategy to Miloš, I memorised his response rather than taking notes in front of him. Despite my best efforts, I must have looked very interested in his reaction because, without any further prompting from me, Miloš gave me a five-minute explanation of the limitations of carbon-dating. As I was surprised by the length of his explanation, I asked Miloš: 'Why do you think it is important for me to know all this?' Miloš responded, now looking slightly irritated, 'Because you asked me before whether I trusted these carbon dates, and I think it's important to show you the methodology to obtain them and all the uncertainties.'

Miloš's reaction to my questioning, and specifically his efforts to show me his competence and reasoning by articulating his scepticism of carbon-dating, led me to rethink my research entirely. In other words, my breaching experiments, initially aimed at disrupting, if only slightly, Rob and Miloš's assumptions, eventually disrupted my own. I wrote in my research diary, 'Why does Miloš use carbon estimates, despite knowing that they have some limitations? Can trust and scepticism exist together? Is Miloš's attitude towards carbon-dating an exception or normal practice?' From that moment, I reoriented my research and sought to answer these questions. I used breaching questions to force Miloš to articulate his background assumptions, particularly his doubts about the accuracy of accepted methods, technologies, ideas or people. At this stage, my research strategy became so obvious that Miloš often anticipated my questions ('I knew you'd ask me about this!'), but he went along with my experiment and answered my questions.

After prompting and documenting similar reactions to Miloš's regarding the use of carbon-dating, I shifted my analysis from being exclusively on trust to the relationship between trust and scepticism in science. I realised that Miloš's seemingly ambivalent use of carbon-dating was not so. I re-read the STS literature that had inspired me to focus initially on the role of trust in science, thinking: 'All these very clever people can't have missed the role of scepticism in science, can they?!' After doing a second review of the literature, I concluded that while early STS scholars conceded that scepticism is part of ordinary practice in science, they also insisted that it is fundamentally dependent on existing trust relationships and background expectations (Ramírez-i-Ollé, 2019b). For instance, in the carbon-dating example, I interpreted that Miloš relied on carbon estimates – even if he knew their limitations –because doing so allowed him to examine and to be sceptical of more important aspects of his work, like crossdating subfossil wood.

At the time I experienced this research turn as an embarrassing realisation that I had not read the literature 'properly' nor had asked the 'right' questions from it. If only I had known that such research turns are perfectly normal in social research! As the anthropologist George Marcus (2009: 22) admits, 'There is not only a tolerance for, but even an expectation of, a shift in plans in fieldwork. This has the standing of a trope in ethnographic writing, a story of "correction" as I call it.' Knowing about the conventions of my academic community in advance – in this case, knowing that my doctoral supervisors expected my research questions to change throughout – would have certainly made the last stages of my research more enjoyable.

At the stage of crossdating measurement series from sub-fossil and living tree samples, Miloš and Rob decided to work together to validate such matches. Up until then, Miloš had worked on his own as he had been Rob's laboratory technician prior to starting his PhD and Rob considered him competent enough to carry out most of the crossdating work alone. At this stage, however, Rob and Miloš worked together because crossdating subfossil chronologies was seen as particularly uncertain and difficult. Rob and Miloš crossdated the chronologies separately, using the same radiocarbon-dating estimates and datasets, and compared the resulting chronologies. Rob had created a web-based spreadsheet where he and Miloš could simultaneously upload and compare the dates of their series (see right-hand column 'Agreement with Miloš?', Figure 2.7). Rob and Miloš's purpose was to find a long and strongly correlated overlap between the so-called 'floating chronologies' from subfossil samples in the Cairngorms (Western part of the Scottish Highlands). Suggestively, the subfossil chronologies are called 'floating' because their dates remained uncertain until more samples were measured and the resulting measurement series were included in the master chronology.

In January 2015, Rob sent an email to the members of the Scottish Pine Project team, announcing that he had finally created an 800-year-long chronology from Scotland. In the last few months of 2014, while Miloš had been finishing his thesis and working on other aspects of the climate reconstruction for Scotland, Rob had tried to crossdate the subfossil samples that had remained undated. For months, Rob had been unable to find a satisfactory overlap between chronologies in the period from the 1440s to the 1540s. 'This is the period of weakest replication and has been a headache for me over the past few months', Rob explained in the email. In order to 'fill the gap' with more samples from the period between the 1440s and 1540s (expressed above as 'weak replication') Rob had decided to focus the 2014 fieldwork expedition on generating more subfossil samples. Rob also explained in the email that the process has been facilitated by the fact that reasonably replicated Scottish BI chronologies 'crossdate' with Jaemtland MXD and Rogen BI data in central Sweden. That is, Rob had interpreted the synchronicity between the provisional 800-year-long chronology of Scotland and two chronologies from Scandinavia created by two members of the team, Björn and Hans, as a confirmation of the dating of the Scottish one (these comparisons are 'Scot-Jaem' and 'Scot-Rogen' in Figure 2.8).

Overall, the creation of the Scottish chronology had taken seven years of immense team work and personal investment from Rob and Miloš in particular. Tree-ring

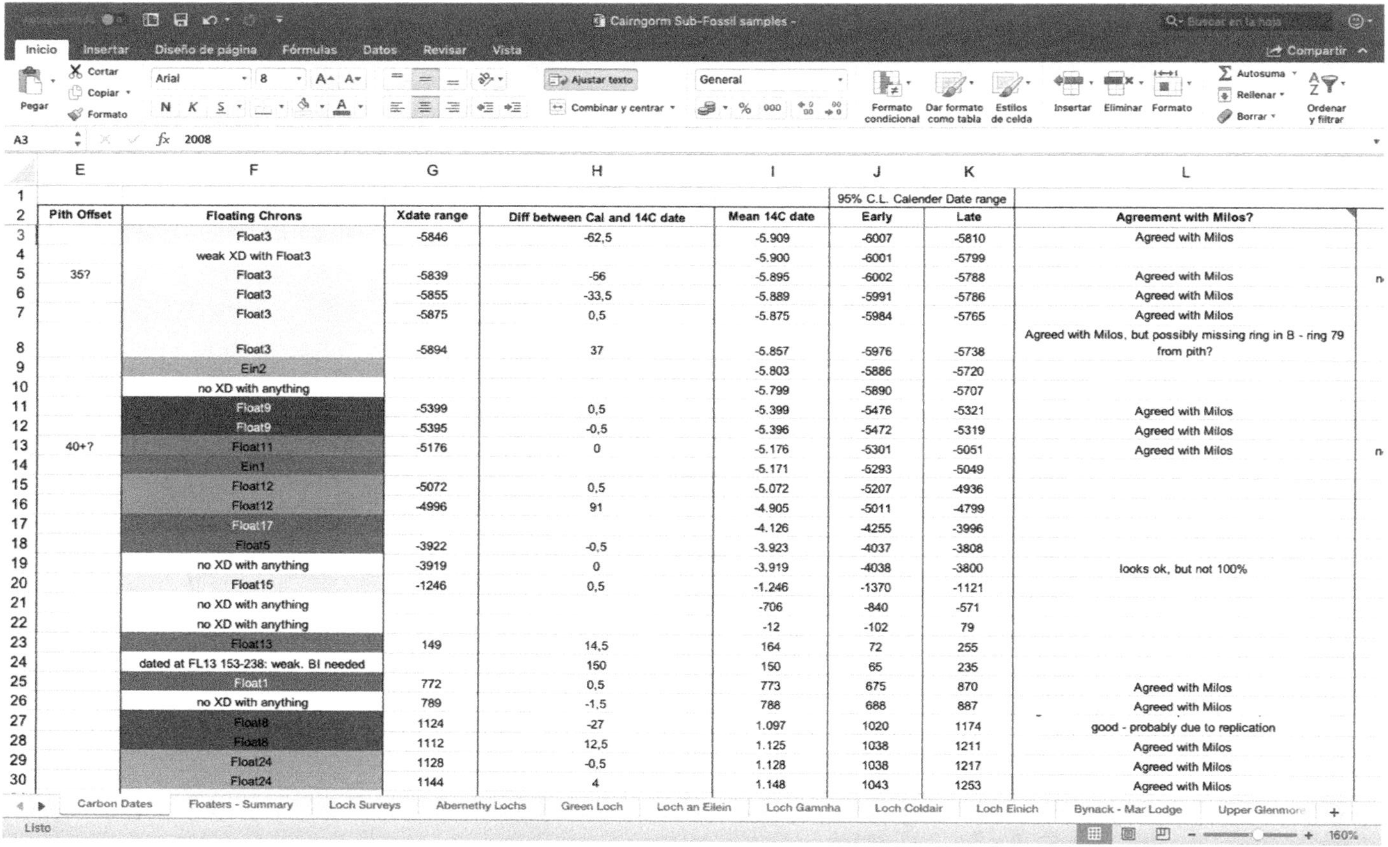

	Pith Offset	Floating Chrons	Xdate range	Diff between Cal and 14C date	Mean 14C date	95% C.L. Calender Date range Early	Late	Agreement with Milos?
3		Float3	-5846	-62,5	-5.909	-6007	-5810	Agreed with Milos
4		weak XD with Float3			-5.900	-6001	-5799	
5	35?	Float3	-5839	-56	-5.895	-6002	-5788	Agreed with Milos
6		Float3	-5855	-33,5	-5.889	-5991	-5786	Agreed with Milos
7		Float3	-5875	0,5	-5.875	-5984	-5765	Agreed with Milos
8		Float3	-5894	37	-5.857	-5976	-5738	Agreed with Milos, but possibly missing ring in B - ring 79 from pith?
9		Ein2			-5.803	-5886	-5720	
10		no XD with anything			-5.799	-5890	-5707	
11		Float9	-5399	0,5	-5.399	-5476	-5321	Agreed with Milos
12		Float9	-5395	-0,5	-5.396	-5472	-5319	Agreed with Milos
13	40+?	Float11	-5176	0	-5.176	-5301	-5051	Agreed with Milos
14		Ein1			-5.171	-5293	-5049	
15		Float12	-5072	0,5	-5.072	-5207	-4936	
16		Float12	-4996	91	-4.905	-5011	-4799	
17		Float17			-4.126	-4255	-3996	
18		Float5	-3922	-0,5	-3.923	-4037	-3808	
19		no XD with anything	-3919	0	-3.919	-4038	-3800	looks ok, but not 100%
20		Float15	-1246	0,5	-1.246	-1370	-1121	
21		no XD with anything			-706	-840	-571	
22		no XD with anything			-12	-102	79	
23		Float13	149	14,5	164	72	255	
24		dated at FL13 153-238: weak. BI needed		150	150	65	235	
25		Float1	772	0,5	773	675	870	Agreed with Milos
26		no XD with anything	789	-1,5	788	688	887	Agreed with Milos
27		Float8	1124	-27	1.097	1020	1174	good - probably due to replication
28		Float8	1112	12,5	1.125	1038	1211	Agreed with Milos
29		Float24	1128	-0,5	1.128	1038	1217	Agreed with Milos
30		Float24	1144	4	1.148	1043	1253	Agreed with Milos

2.7 Rob and Miloš negotiating online the accuracy of crossdating subfossil data.

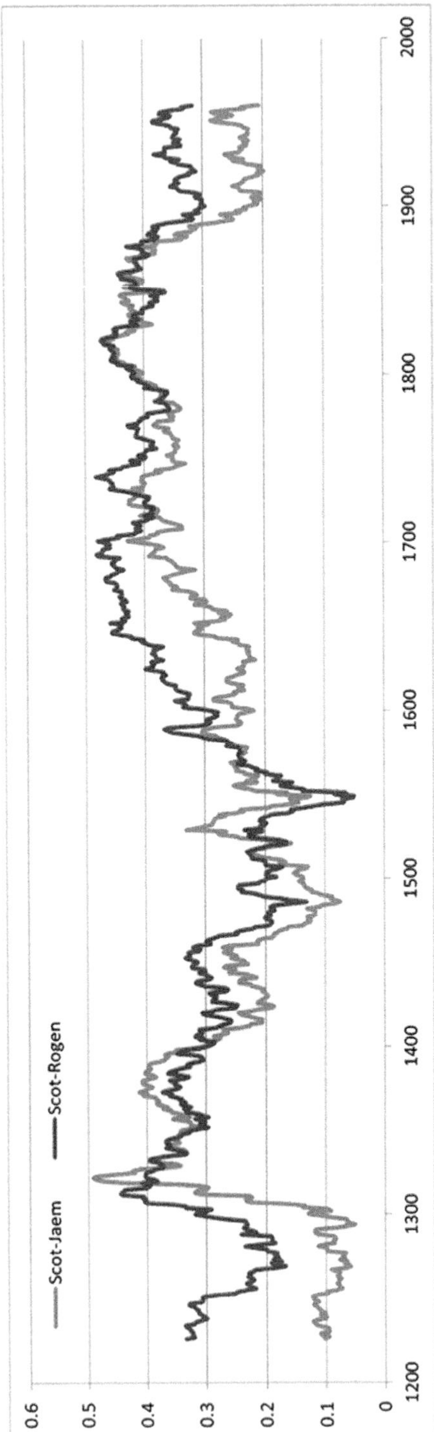

2.8 Rob validated the accuracy of the 800-year-long chronology of Scotland by noting its similarity with two chronologies from Sweden created by team members.

chronologies are often colloquially named after the region and the individual dendrochronologist who presumably undertakes the largest share of work (for instance, 'Douglass's Aztec-Pueblo Bonito chronology', 'Baillie's Irish oak chronology' and the 'Schweingruber network'). In conferences, I occasionally heard dendrochronologists talking about '[Rob] Wilson's Scottish chronology', which means that the 800-year-long chronology (officially named in publications as 'N-Cairn' – Northern Cairngorms) was considered to be part of the shared resources upon which others could draw to validate their chronologies (in the same way that Rob relied on Björn and Hans's Scandinavian chronologies).

Conclusion

This chapter has described the work of producing tree-ring data from the wood samples generated during fieldwork. The creation of the 800-year-long tree-ring chronology from Scotland was a huge achievement for Rob and the Scottish Pine Project team because they managed to extend further back in time the existing chronology created by Malcolm Hughes in the 1980s that started in 1721. Dendrochronologists base part of their reputation on creating new tree-ring chronologies and updating and extending old ones, and their professional identity is intimately tied to such epistemic objects. For this reason, when Rob met his colleagues in conferences, one of the most recurrent questions he received was 'How far are you back now?', in reference to the exact length of the chronology. For dendroclimatologists like Rob and Miloš, the length of the chronology is also of immediate importance because, as we shall see in Chapter 4, it determines the temporal scope of their claims about historical climate change.

The three stages that I have identified as characteristic of the work of dendrochronology carried out in Scotland (reading and representing tree-rings and synchronising tree-ring data) were driven by the firm belief that the dating of tree-rings, and specifically the matching of ring patterns across different tree samples, was possible (but not certain) to achieve. Miloš's work of detection and correction of 'mismatches' between tree-ring measurements – by inferring the existence of missing and false rings – was based on the assumption that pattern-matching was the normal and expected outcome given that the sampled trees had been affected by similar environmental conditions. Of course, dendrochronologists also encounter cases of 'uncrossdatable' samples, as is the case with some of the Alladale samples in Scotland or samples from tropical trees that have very indistinct ring boundaries (significantly, the exceptionality of the crossdating work done with tropical samples is qualified as 'frontiers of dendrochronology' in the most recent textbook; Speer, 2010: 253). In this way, dendrochronologists routinely adjust their expectation of finding common tree-ring patterns to the specific empirical qualities and the structure of the wood under study.

Dendrochronologists' expectation of achieving crossdating has grown over time as they have documented consistently similar ring patterns across tree species around the world. The acknowledged founder of dendrochronology, Andrew

Douglass, explained how he first identified in the early 1900s the phenomenon of crossdating in Northern Arizonian forests:

> Our first experience is so symbolical of later application that it is recounted here. In January 1904 the uniformity around the circuit of a tree was observed and recorded. A few months later five sections newly cut near Flagstaff were carefully measured and a group of small rings was noted some 21 years in from the bark. While this was fresh in mind, curiosity raised the question whether this same group could be found near the outside of a slightly weathered stump whose date of cutting was unknown. On examining the rings in the stump the group was found at once but was only 11 rings in from the outside. So that tree must have been cut in 1894. The owner of the land was sought and asked when his timber was cut and he answered, in 1894. This incident supplied the mechanical key by which many tree-ring problems have been solved'. (Douglass, 1937: 3)

More than seventy years after Douglass's discovery, two reputed dendrochronologists claimed that

> Admittedly, crossdating is not universal among all tree species. However, its occurrence over a broad range of taxa [tree species] growing in extremely diverse habitats worldwide indicates that crossdating is a property of tree growth that frequently *emerges* from the radial growth increments of trees being subjected to a highly variable set of microenvironmental conditions. (Cook and Pederson, 2011: 83–84)

In other words: while in the early twentieth century crossdating was thought to be geographically circumscribed to one forest, by the early twenty-first century, it was known to be an almost universal feature of all trees.

Dendrochronologists' empirical confirmation and continued success in developing longer and more heavily corroborated chronologies has reinforced their trust not only in the predictive value of crossdating (hence, why it is described as a 'principle'; Speer, 2010: 11; Fritts, 1976: 20), but also in the fundamental assumption upon which dendroclimatology is based, namely that trees are appropriate sources of climatic information. As stated by Harold Fritts (1976: 21), 'the fact that crossdating can be obtained itself is evidence that there is some climatic or environmental information common to the sampled trees'. Similarly, Malcolm Hughes argued (2011: 31) that 'So far as a cause for these common patterns is concerned, the prime suspect is climate variability. So, in turn, where tree-ring samples "cross-date" (share massively replicated patterns of variability), a prima facie case for their containing a climate signal has been made.'

The robustness of the 'principle of crossdating' as a foundational practice of dendrochronology stems from the fact that it is continuously and collectively verified every time dendrochronologists create or update existing tree-ring chronologies. Miloš's initial work of 'chronology-building' among the individual datasets from Scotland (including the online work with Rob) and Rob's later comparison of the 800-year-long chronology from Scotland against the chronologies from Sweden are illustrative of the way the evidence of the principle of crossdating grows from the entanglement of peers who borrow and validate each other's chronologies, and in this way reinforce ties of personal trust and shared knowledge.

The 'principle of crossdating', I suggest, should be regarded as a 'social institution'. We often think of institutions as something physical like a building or an organisation but they can also be an intangible idea or social norm like the institution of 'marriage', 'slavery' or 'freedom of speech' that result from the referential and normative activities of a collective (see Brisset, 2017; Muniesa, 2014; Bloor, 2013; Rafanell, 2009; Schyfter, 2009; MacKenzie et al., 2007; MacKenzie, 2006; Callon, 1998; Barnes, 1983 for an extended account of the 'performative theory of social institutions'). On the one hand, the principle of crossdating results from referential activities in the sense that its existence depends upon individuals like Andrew Douglass first referring to an external and observable reality (trees in Northern Arizona) as 'crossdated', and subsequent generations of dendrochronologists invoking such authoritative references as models for their work of recognising an unknown empirical reality (new tree species in new sites) as 'crossdated'. On the other hand, the principle of crossdating is the result of normative activities in the sense that the work of crossdating is subject to collective standards and evaluations of correct and incorrect application. For instance, as we have seen, Andrew Douglass warned that 'the operator does not dare to seek relief from his responsibility' and Miloš insisted that 'you cannot allow statistics to manipulate you'. More generally, dendrochronologists' insistence on differentiating their work from the improper ring-counting work of foresters is also indicative of the normative aspects that characterise crossdating as a social institution, which in turn serve the larger purpose of establishing the professional identity and the methods of dendrochronologists as 'scientific'.

The practice of crossdating – institutionalised through collective processes of reference and correction among dendrochronologists – has changed over time due to the changing needs of dendrochronologists and the greater diversity of the tree-ring patterns studied. I cannot emphasise enough the idea that the dendrochronology work described in this chapter is specific to Rob and Miloš's working environment. This obvious point became clear to me after reading in the main dendrochronology textbooks (written by dendrochronologists from the United States) that 'skeleton-plotting' was a common crossdating technique. This method consists of plotting by hand the relative length of the rings of each tree sample onto graph paper and comparing measurements by sliding one graph paper past another (Speer, 2010: 11). Shortly after starting work as a technician, I realised that the work we did in Scotland did not resemble what I had read about skeleton-plotting and I, confused, asked Miloš when we would start crossdating. He looked at me, surprised, and said: 'We are already crossdating!' Miloš explained that, unlike in the Southern USA where skeleton-plotting had been developed, trees in Scotland did not produce as sensitive tree-rings. 'As you know, we need to use microscopes to see some of the narrow rings that trees produce here because it is cold and wet.' Miloš added, laughing, 'Also, if I had to draw a plot for all the samples we have, I would not be able to finish my PhD in three years!' In this way, the use of computerised and statistical methods for crossdating allowed Miloš to work with the Scottish samples and to produce the amount of new tree-ring data required to obtain a doctorate.

My conclusion that dendrochronology is sustained through local acts of reinterpretation and empirical examination of an institutionalised practice speaks to public criticism of climate science as being the result of tribalism, groupthink and dogmatism. In the aftermath of the hacking of climate science emails, newspapers from the whole ideological spectrum published editorials warning climate scientists against any further 'temptation to fall into tribalism'. A *Guardian* editorial called 'Truth and tribalism' argued that 'it is true that many of the specific sins involved, such as partial peer-reviewing and overly zealous defence of one's own research, are and always have been found in all manner of science departments. With climate, though, the stakes are higher – and so the standards must be too' (The Guardian, 2010b). An editorial from the *Wall Street Journal* (2009) criticised climate scientists by quoting Professor of Human Geography Mike Hulme saying, 'The tribalism that some of the leaked emails display is something more usually associated with social organization within primitive cultures; it is not attractive when we find it at work inside science.' Many bloggers used religious metaphors to criticise climate scientists as 'believers' and to characterise climate science as 'dogma' (Nerlich, 2010). In response, a group of 255 scientists, all members of the US National Academy of Sciences, also employed the language of 'dogma' to rebuff their critics as 'Many recent assaults on climate science and, more disturbingly, on climate scientists by climate change deniers, are typically driven by special interests or dogma, not by an honest effort to provide an alternative theory that credibly satisfies the evidence' (Ramírez-i-Ollé, 2015a: 393; The Guardian, 2010a).

Anti-authoritarianism is an ideal firmly established in the minds and hearts of scientists, as seen in the fact that the Royal Society's motto is 'Nullius in verba', which means 'Take nobody's word for it'. The ideal that scientists do not obey any authority also underlies popular perceptions of historical conflicts between science and religion (Numbers, 2009 shows why such a 'conflict myth' is false). While scientific practice is depicted as driven by unconditional adversarialism and untainted scepticism, religious practice is portrayed as inherently dogmatic and fanatical (see Berger and Zijderveld, 2009 on why the latter is incorrect). This false dichotomy is echoed by the physicist Richard Feynman (1998: 43), who argued that 'The uncertainty that is necessary in order to appreciate nature is not easily correlated with the feeling of certainty in faith, which is usually associated with deep religious belief. I do not believe that the scientist can have that same certainty of faith that very deeply religious people have.'

In response to those who criticise climate scientists for conforming to the traditional beliefs and practices upheld by their community, I offer this chapter as counter-evidence. If dendrochronologists seek to identify dating errors and false and missing rings in a chronology, how can they *not* expect the rings to match and the principle of crossdating, empirically verified by hundreds of colleagues before them, to apply to the case at hand? Given that Rob and Miloš did not have the resources or the expertise to carbon date all the samples, is it not perfectly rational to defer to other experts and their established methods? If Rob sought to confirm the dating of the 800-year-long chronology against reliable chronologies, why would he dispense with the work of two trusted team members? More generally, if a scientist seeks to critically examine the result of an experiment, how can he or she do so if not by taking certain aspects of the experimental set-up (e.g. the reliability of instruments, the measurement system) for granted (Collins, 1985 describes the problems that scientists potentially face when distrusting all the aspects of a replication as 'experimenters's regress')? Why do we expect scientists to be more disloyal to their 'tribe' of peers than other social groups? How can someone aspire to have his or her work recognised by relevant colleagues if he or she systematically distrusts them to offer sound advice?

The historian of science Thomas Kuhn famously argued in an article called 'The function of dogma in scientific research' that received culture, rather than hindering the progress of science, provides its basis. Kuhn (1963: 348) started with the historical observation that

> From Galileo's reception of Kepler's research to Nägeli's reception of Mendel's, from Dalton's rejection of Gay Lussac's results to Kelvin's rejection of Maxwell's, unexpected novelties of fact and theory have characteristically been resisted and have often been rejected by many of the most creative members of the professional scientific community [...] Preconception and resistance seem the rule rather than the exception in mature scientific development.

For Kuhn (1963: 359), the function of dogma is to facilitate a focused investigation:

> The practitioners of a mature scientific specialty are deeply committed to some one paradigm-based way of regarding and investigating nature. Their paradigm tells them about the sorts of entities with which the universe is populated and about the way the members of that population behave; in addition, it informs them of the questions that may legitimately be asked about nature and of the techniques that can properly be used in the search for answers to them.

As Barry Barnes (1982: 19) also expressed it, dogma has both a cognitive and social function in the sense that 'dogmatic training effectively and beneficially binds the scientist to his fellows, so it effectively and beneficially relates him to nature. Nature is too complex for random, unsystematic, diffuse investigation.'

Given that research in a scientific field can only proceed – as exemplified in this chapter with the creation of a longer tree-ring chronology – because its members share a particular viewpoint of the world, and such standardised cognition facilitates communication and mutual examinations of labour, I find it profoundly unfair to criticise climate scientists for being dogmatic on certain issues. As I illustrate in Chapter 5, climate scientists from across different subfields share some dogmas but they also disagree on fundamental issues; in this way, while dendroclimatologists regard the principle of crossdating as incontrovertibly true, some paleo-modellers think it is utterly inadequate.

3

Standardisation

As it was, the 800-year-long chronology from Scotland that Rob and Miloš had painstakingly created did not yet serve their interests as dendroclimatologists. Rob and Miloš were well aware that other factors besides climate could affect tree-ring growth, and that these confounding factors were reflected in the ring-width and blue intensity chronologies. Identifying such non-climatic factors and removing their trend from tree-ring chronologies is a practice described in textbooks as 'standardisation' or 'detrending'. This practice takes its name from the act of transforming series of tree-growth measurements into 'standard' tree-growth indices, calculated as deviations from a mathematical model of expected tree-ring growth called a 'standardisation curve'. The residuals from this curve are taken to represent the effects of climate on tree growth.

Since the 1980s, dendrochronologists have employed a conceptual model called the 'linear aggregate tree-growth model' in order to standardise tree-ring data and to distinguish between the different factors affecting tree-growth. I first learned about this model while attending Rob's dendroclimatology class, suggestively called 'Seeing through the forests – teasing out climatic information from trees' (see power point slide, Figure 3.1). The creator of the linear aggregate tree-growth model, Dr Edward Cook, developed it as part of his doctorate (Cook, 1985) building upon the so-called 'Liebig's law of the minimum'. This law is used in dendroclimatology as a 'first approximation of the environmental factor that is most likely to be recorded in a given tree-ring chronology' (Speer, 2010: 15). Besides the effects of 'climate' (C), Cook also identified three other factors that were recorded in any given tree-ring chronology: the 'age' (A) of trees, measurement errors (E) and two types of 'disturbance' (D); one form originating within the forest itself, like a blowdown of trees (D1), and another originating outside the forest often related to fires, pollution or logging (D2).

Dendrochronologists describe the practice of standardisation using the engineering metaphor of 'noise reduction'. The concept of 'noise' is used to refer to any kind of undesired interference in the transmission of valuable information or a 'signal' (this metaphor has also been used to analyse other social activities such as sports, finance and card games; see Silver, 2012). For dendroclimatologists like Miloš and Rob, 'reducing noise' means eliminating those tree-growth trends

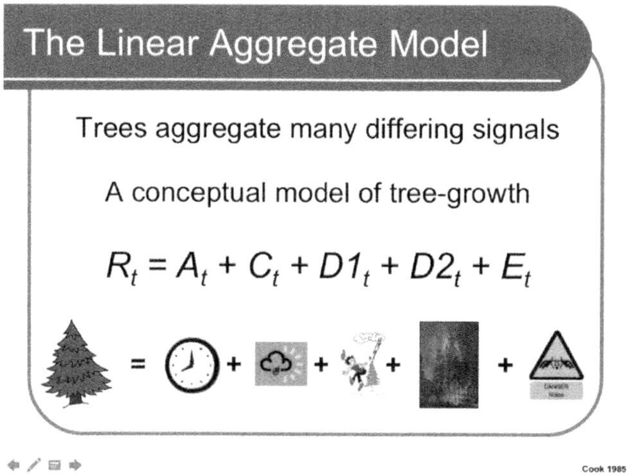

3.1 A conceptual tool for distinguishing the 'climatic signal' in tree-ring data.

conceptualised as distinct from the climatic signal. However, as Cook (1987: 38) explains himself, what counts as 'signal' and 'noise' in his model is a matter of perspective, as 'one researcher's "signal" would frequently be another researcher's "noise"'. While for a dendroclimatologist, 'climate' is the signal and 'disturbance' is the noise, for a dendroecologist who is interested in the study of forest dynamics and ecology, 'disturbance' may be instead the signal of interest and 'climate' is not as relevant.

Despite the fact that 'noise' and 'signal' are not fixed attributes of tree-ring data, but rather their attribution depends on the user, all dendrochronologists agree that the biological age trend is a noisy factor that needs to be 'cleaned' off from the dataset. Consequently, standardisation has traditionally been associated with the process of eliminating the ageing trend from the tree-ring chronologies. The most used statistical package, called 'AutoRegressive Standardisation' (ARSTAN), used for modelling and removing the effect of ageing (also created by Edward Cook) includes two types of 'standardisation curve' or standardisation approaches. The first group consists of deterministic mathematical functions or 'negative exponential curves' that assume that ring width normally decreases from pith to bark over the life of a tree, as older trees cannot produce rings as wide as when they were young. The second group consists of non-deterministic statistical functions known as 'smoothing splines' that do not entail any a priori assumption about the ageing trend (either downward, upward or linear).

The practice of standardisation essentially involves developing the necessary judgement to distinguish which aspects of the trend in the tree-ring chronology reflect the effect of 'noise' (age) and the resulting desired 'signal' (either climate or disturbance). The risk for dendroclimatologists is choosing a standardisation curve that not only removes the ageing trend but, inadvertently, also eliminates the long-term growth trend related to changes in precipitation or temperature (climate). All dendroclimatologists I talked to experienced standardisation as one of the most challenging aspects of their work. At an international conference in Melbourne, many presenters consistently finished their talks with references to the uncertainties of choosing a particular standardisation approach. In one plenary session, the speaker was asked 'how to resolve the problem of choosing one standardisation curve'. In one textbook, Edward Cook and another reputed dendroclimatologist, Dr Keith Briffa, acknowledged that the choice of standardisation curve is a difficult one, which has consequences for the resulting chronology and climate reconstruction. However, they refused to propose any guidelines for choosing a standardisation curve, because 'this decision is likely to be completely data and application dependent' (Cook and Briffa, 1990: 161). Cook and Briffa also insisted that dendroclimatologists should 'never use any tree-ring standardization method or computer program as a *black box*' (1990: 161).

The perceived arbitrariness in choosing a standardisation curve has been the focus of public criticism. One blogger criticised Cook and Dr Briffa 'because in few if any of [their] papers is it made clear why this choice was made, and there is no set of agreed-upon rules or guidelines anywhere that clearly defend and delineate under what conditions the ICS [Individual Curve Standardization] method should and should not be used to detrend tree ring series' (Bouldin, 2012). Another blogger called Steve McIntyre – the creator of the popular blog 'Climate Audit', which, as its name indicates, is set to audit climate research (Sharman, 2014) – has conducted extensive and detailed analyses of the standardisation procedures of dendroclimatologists, particularly of Cook and Briffa's work (McIntyre, 2016, 2009).

Throughout my entire research, Miloš and Rob always insisted that deriving climatic information from trees growing in Scotland was particularly difficult due to the effects of past human disturbance on tree growth. 'If we manage to do dendroclimatology in Scotland', Rob stated in front of an audience of a hundred colleagues at the conference in Melbourne, 'We will make a case for our field.' Rob still nowadays considers the Scottish Pine Project the most challenging dendroclimatological project of his career: 'The behaviour of trees in British Columbia, where I did my Masters, was pretty crystal clear.' Miloš also regarded the experience he was gaining while working in Scotland as an advantage: 'Everything else I find in the future can only be easier!' Rob and Miloš complained that doing dendroclimatology in Scotland was 'complicated' because they found it difficult to interpret the patterns of some tree-ring chronologies. Over time, they came to suspect that the dataset from Scotland contained 'noisy' factors (other than age), which they had to identify and remove.

The 'epistemological conundrum' that Rob and Miloš faced at this stage was, therefore, to identify and minimise, as far as possible, unwanted non-climatic influences, and to bring to the fore those changes in tree-ring growth that could be attributed to changes in temperature alone. I characterise Rob and Miloš's work of standardisation as involving two stages: sorting out and eliminating noise. As I describe below, and largely as a result of the specificities of the Scottish environment and tree-ring patterns, Rob and Miloš faced difficulties in applying traditional standardisation techniques. As a result, they had to find new ways of standardising the Scottish data and rendering these new methods and choices credible to colleagues and outsiders.

Sorting out noises

In 2007, Rob noticed that 'something was off' with the ring-width data from Scotland. What Rob meant is that the ring-width chronologies from Glen Affric, a site in the North-Western region of the Scottish Highlands, did not show the declining trend that he expected to find due to the ageing of trees. Instead, the ring-width chronologies showed irregular peaks of growth around the decade 1820–1830, which were not related to any documented increase of temperature in Scotland. Rob did not have any reason to suspect that the irregular fluctuations in the ring-width data were due to any measurement errors, as he thought of the student who generated the data as an 'excellent lab worker'. Rob also knew that the tree-ring data that Malcolm Hughes had used to create the first temperature reconstruction in Scotland also showed the same growth peaks. Therefore, Rob suspected that the abnormal ring-width patterns from Glen Affric were related to 'disturbance' events, but he did not know exactly what these were.

From 2008 to 2012, Rob and the members of the Scottish Pine Project fieldwork expedition sampled other locations throughout the Scottish Highlands with the aim of investigating, among other issues, the spatial extent of the disturbance first identified in the chronologies from Glen Affric. In one report that Rob prepared in 2008 for the landowners of the Rothiemurchus Estate in exchange for being granted easier access to the sites (Chapter 1), Rob explained that 'tree-ring data were also utilised from various other sites throughout the Highlands to determine the applicability of the results from Glen Affric to the remainder of the country' (Wilson, 2008). Rob soon discovered that many of the new ring-width chronologies, especially from the West of the Scottish Highlands, showed similar irregular patterns for the years 1820–1830.

Rob sought to explain the disturbances by familiarising himself with the ecological and social history of Scotland. He talked to the government foresters who had facilitated access to the fieldwork site (Chapter 1) and read books written by historians and ecologists that documented the occurrence of severe storms and logging events in the Scottish Highlands for the period 1820–1830, exactly the period when Rob had observed disturbances in the Scottish data. These sources also reported that pine woodlands in the Scottish Highlands had suffered consecutive thinning and clear-cutting from the sixteenth century onwards; and the period of greatest activity was the nineteenth century, due to the Napoleonic Wars (1803–1815), when timber was extracted for economic and war efforts. On this basis, Rob named the disturbances he observed in the Scottish data as 'Napoleonic Impact Bias'.

Leah, the dendroarchaeologist of the Scottish Pine Project, also assisted Rob in further identifying the nature of this disturbance. Leah provided Rob with information about changes in the timber supply and woodland resources in Scotland through the analysis of historical documents such as diaries and official documents. Like Rob, Leah was interested in locating all the remaining Scots pine-grown pinewoods of Scotland and generating a tree-ring chronology with material from archaeological buildings. However, their collaboration was established upon a different definition of the noisy aspects of tree-ring data. In fact, Rob often dismissed jokingly as 'shite' the effect of wars and population changes on ring-width data, which was Leah's interest as a dendroarchaeologist. Leah was well aware of the divergence of interests with Rob, which she thought suited both parties. 'I am happy to work with Rob's "leftovers"', Leah told me, laughing.

In early 2012, Rob started investigating the potential wider presence of human disturbance in the Scottish dataset. He analysed two datasets from the Cairngorms, the Eastern area of the Scottish Highlands. He wanted to compare ring-width chronologies from relatively lightly managed woodland sites (Loch an Eilein and Creag Fhiaclach) against chronologies from a more highly impacted woodland (Loch Gamhna). Rob hypothesised that, if disturbance and the effects of logging and forestry management indeed existed in the Loch Gamhna dataset, he would observe distinctively wider ring-width patterns. This is because, he reasoned, the removal of trees in a forest decreases competition between trees for light and nutrients and allows the remaining trees to grow faster. Rob delegated the execution of this experiment to an undergraduate student who conducted this work as part of her dissertation. This student created a graph that showed how the ring-width chronologies from Loch Gamhna deviated from the disturbance-free chronologies (Loch an Eilein and Creag Fhiaclach) in three periods, which coincided with forest management events in the Rothiemurchus Estate that Leah had previously identified (Figure 3.2).

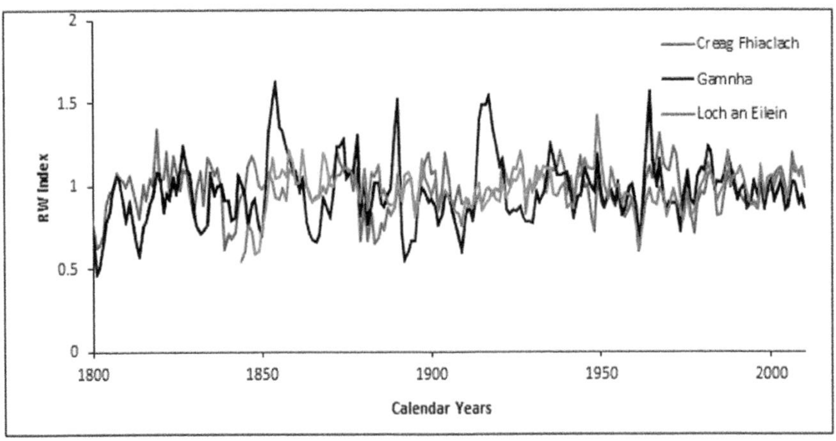

3.2 The first confirmation of disturbance in the Scottish dataset.

After confirming the presence of human disturbance at one sampling area in the Cairngorms, Miloš proposed employing a tree-ring model to investigate the existence of disturbance in the wider dataset. Modelling was, and still is, a relatively recent method used in dendroclimatology to explore tree-ring data. In particular, Miloš employed a model of tree-ring-width formation called 'VS-Lite' to generate a hypothetical picture of what a ring-width chronology would look like if only the effects of climate in Scotland affected tree growth. Miloš's aim was to generate 'ideal' modelled chronologies against which to compare the 'real' Scottish data, so as to confirm the presence of disturbance. 'Any differences between the modelled and observed ring-width data', Miloš explained, 'can only be attributed to disturbance.'

In August 2012, Rob arranged for Miloš to visit the Lamont laboratory in New York, where one of the main creators of the VS-Lite model, Dr Kevin Anchukaitis, was working. Rob personally knew Kevin and had worked with him (in 2012, Rob and Kevin were working together on an article published as Anchukaitis et al., 2012; see Chapter 5). In addition, Rob had been a postdoctoral researcher at Lamont and maintained an affiliation as an adjunct scientist (this is the same laboratory where Rob stored his supervisor's samples; Chapter 2). At the time of writing this text, Edward Cook was still the director of Lamont; an institution with long reaches into the climate research community as it had become a 'centre of pilgrimage' for many dendroclimatologists (including myself, see Box 5).

Box 5

How well do I need to understand an epistemic object to explain its formation?

By the end of my first year of fieldwork (April 2013), Edward Cook's reputation as a researcher and teacher had reached my ears and I was keen to meet him. I also wanted to learn more about the standardisation techniques that Cook had developed, which Rob and Miloš used extensively. At this stage of my research, I could barely observe how Miloš standardised the data because he often worked very late at night from his house, and was, at times, away from Edinburgh. Consequently, my three-year-long fieldwork evolved from intense physical 'co-location' – working in the laboratory in St Andrews once per week, joining the yearly expeditions in the Scottish Highlands, and attending weekly lectures, workshops and conferences – to various forms of 'co-presence' (Beaulieu, 2010), which included keeping in touch by email and, whenever we could, meeting in St Andrews and Edinburgh.

Partly as a result of the changing nature of my fieldwork, and partly because of my own perceived lack of training in statistics, I felt I had an insufficient understanding of the 'nitty-gritty' of standardisation. I could have adopted, as Latour and Woolgar ([1979] 1986: 43) propose, a 'naïve approach' in the study of cultures, which would have meant remaining ignorant of standardisation. While I played the role of the 'stranger' throughout my fieldwork (with the use of breaching experiments; Box 4), it became obvious that it would be impossible, and possibly detrimental to my own research relations with Rob and Miloš, not to seek to understand, as an 'insider', the technical vocabulary and practices of dendroclimatologists. As a result, in January 2014 I joined a course on 'statistical dendroclimatology' in Tasmania led by Cook himself, which took place before a major dendrochronology conference in Melbourne (Australia) that I was also attending (Box 2). This statistics course was part of a broader training space called 'fieldweek', whereby experienced dendrochronologists taught different aspects of dendrochronology and cohabited with students for a week in a remote location. Quite understandably, these fieldweeks were very popular among dendrochronology students (can you imagine spending a week with your favourite sociologist, anthropologist or historian?), and I wanted to study these spaces of socialisation. As an overt epistemographer, I worked with my colleagues in the statistics group on completing assignments and presentations while also conducting interviews with many of the fieldweek participants, particularly Cook, my teacher.

After participating in the fieldweek on statistical dendroclimatology, I had certainly learnt a great deal about standardising tree-ring data and I could even perform a few basic calculations, but what exactly had changed in terms of my improved ability to conduct my epistemography? First, I realised how innovative Rob and Miloš's standardisation work was, and how specific their challenges were, in comparison with past and existing work done by their colleagues. Secondly, through conversations with Cook, I learnt about the history of the different standardisation curves and programmes used by Rob and Miloš. While this background knowledge has not been directly relevant for writing this chapter, it has provided me with a greater awareness of the potential open-endedness of Rob and Miloš's standardisation practices. Lastly, and perhaps more importantly, after participating in the fieldweek, I gained acceptance from many expert dendroclimatologists, as in their opinion I had achieved sufficient skills in standardisation. Overall, throughout the fieldweek, I and my relations with my subjects of study experienced a transformation that is perfectly described by the sociologist Harry Collins below:

> As a sociologist of science you essay research on a new specialism and you initially understand neither the banter nor the technical terms. After a painful period, if you are lucky you begin to pick up on the inferences in others' conversations and eventually you begin to be able to join in. One day a respondent might say in response to one of your technical queries 'I had not thought about that', and pause before giving you an answer. When this stage is reached respondents will start to be happy to talk to you about physics and even respond generously and with consideration to your critical comments. Eventually people will become interested in what you know, not as a scientist in your own right, but as a person who is able to convey the scientific thoughts and activities of others. If you've just come

> from visiting scientist X you may be able to tell scientist Y something of the science that X is doing … What were once 'interviews' then become 'conversations' that can be interesting and occasionally even useful to both parties. What also happens in a conversation is that by occasionally anticipating a point your partner is about to make you can speed things along. You might also verbally fill in some gaps that might otherwise be forgotten. You can recognize jokes, irony and when you are having your leg pulled (though, in the nature of things, interactional competence does not allow you to recognize lies). When you get good at it you can even take the devil's advocate position in respect of some scientific controversy and maintain it well enough to make your conversational partner think hard. (Collins, 2004a: 104)

I stopped worrying about 'how much' I needed to know about standardisation when, as a result of participating in the fieldweek course, my subjects recognised me as a 'quasi-expert'. Examples of such moments of recognition resemble those recounted above by Harry Collins: when Rob, Miloš and other fieldweek participants occasionally told me, in response to one of my questions, 'I had not thought about that'; when they insisted on talking to me about their work; when, upon Rob and Miloš's request, I was able to tell them something of the science I had learnt in the fieldweek; and when I made everybody laugh during a fieldweek presentation when I joked about the innumerable times that I and colleagues in the statistics group had pressed the button 'ENTER' in the ARSTAN programme. Overall, my 'expertise' as an epistemographer resulted from assimilating and displaying specific skills, vocabulary and conduct that my subjects recognised as relevant for understanding standardisation (see Coopmans and Button, 2014).

As Laudel and Gläser (2007) argue, each research encounter implies a negotiation of the appropriate level of knowledge and technical depth shared between the researcher and subjects, which often involves the co-construction of a simplified language or 'ad hoc-pidgin'. Needless to say, it takes a huge amount of time and effort to develop the necessary field relations and vocabulary to understand and write about the formation of any epistemic object. As Laudel and Gläser (2007: 108) put it, 'There are limits to a scientific preparation when one has to interview a molecular biologist on Monday, a solid state physicist on Tuesday, an electrical engineer on Wednesday and a physical chemist on Thursday.' One solution to the problem of developing a sufficient understanding of any given culture is that a member of such a culture conducts the epistemography itself. There are multiple cases of natural scientists and engineers becoming epistemographers (see Laudel and Gläser, 2007: 96). The challenge for these epistemographers is not a lack of recognition from subjects – as is the case with epistemographers like myself turning into 'quasi' natural scientists – but an excess of it. Subjects might not share their ideas with the epistemographer precisely because they think their ex-colleague already knows them.

Upon his return from Lamont, Miloš was able to modify the source code and the parameters of the VS-Lite model. Essentially, the VS-Lite simulates the effects that input precipitation and temperature data might have on tree-ring growth given a set of twelve adjustable parameters. For instance, one of the default parameters of the VS-Lite model was that trees do not grow below freezing temperatures and reach an optimal rate of growth at a defined temperature such as 20 degrees centigrade. Miloš insisted that a competent use of VS-Lite involves adjusting the parameters in a way that makes sense in terms of what the researcher knows about the specific ecology of an area and the general physiology of trees. 'I use the parameters as a starting point', he said. 'You can experiment, but you have to have a justification for the values you use. I mean, I can't set the upper temperature limit to 50 degrees or this lower limit to minus 22 degrees, because it does not make any sense on how trees grow anywhere.'

On the basis of the background knowledge of the ecology of the Scottish Highlands he had gained during fieldwork (see Conclusions in Chapter 1), Miloš decided to experiment with the parameters regarding soil moisture in order to accurately reflect the particularly wet conditions in which trees grow in Scotland, especially in the West. In particular, he included an 'upper soil moisture' threshold to simulate how high amounts of water in the soil might limit tree growth. When I asked Miloš how he had come up with the idea of including a new parameter in the model, he told me that 'I just imagined that the wetness we saw in the field could have an effect on how trees grow in Scotland'. With his VS-Lite model adjusted to the environmental conditions in Scotland, Miloš created two averaged ring-width chronologies for the West and the East of Scotland and compared them against the respective 'real' chronologies. The results indicated that the simulated and the real chronology from the West were more different than those from the East, which Miloš interpreted as an indication that disturbance was more prevalent in the Western sites like Glen Affric, which could be due to high levels of precipitation and the effect of logging. Both Rob and Miloš regarded these results (as well as the crossdating problems that Miloš had encountered with chronologies from the Western site of Alladale; Chapter 2) as a confirmation that noise was present in the broader Scottish dataset. By the end of summer 2012, Rob and Miloš had decided to spend some time working with a new method that they hoped could help them to eliminate the disturbance trends in the Scottish data.

Cleaning out noises

In June 2012 (two months before Miloš visited the Lamont laboratory) Rob attended a workshop where he explained to his colleagues the problem of disturbance they experienced in Scotland. Edward Cook, who was also attending the workshop, mentioned that he was working on a paper with a dendroecologist (Dr Daniel Druckenbrod) about a method that Cook described, as Rob remembered it, as 'magic'. The method, originally called the 'Combined Curve and Trend intervention approach', or 'CCT', (in recent publications, this method is referred to as 'CID' or 'Curve Intervention Detection'; see Rydval et al., 2018) was partly based upon

Cook's ARSTAN standardisation software and it sought to estimate the effect of disturbance on ring growth. As a dendroecologist, Dan's interest was opposite that of Rob and Miloš in the sense that he was interested in disturbance and forest ecology as a 'signal'. However, Cook encouraged Rob to contact Dan and experiment with the CCT method as Cook thought it could be used to identify and later remove series of disturbance indices.

When Miloš visited Lamont in August 2012 to learn about VS-Lite from Kevin, he fortuitously met Dan, who was preparing the article about CCT with Cook. Miloš explained to Dan the disturbance problem in Scotland and they agreed to collaborate. Miloš told me that Dan was pleased to start exploring 'new applications' for CCT and to develop a 'more refined' version with Miloš's help. When I asked Miloš how important the meeting with Dan was, he said, 'The good thing about it was that I had a chance to talk to Dan face to face, which is always helpful if you want to start working with someone.' Miloš clarified that, at the time, there were no published articles about CCT, and he learned how to use it with some guidance from Dan.

After returning from his visit to the Lamont laboratory, Miloš did a few tests that compared the corrected or 'cleaned' chronologies with CCT against the modelled chronologies produced by VS-Lite, which simulated how chronologies would look without disturbance. Miloš reported that the tree-ring chronologies from the West of Scotland needed to be more corrected with CCT than those from the East in order to agree with the VS-Lite chronologies, which he interpreted as an indication that the Western dataset was more extensively affected by disturbance. In an email conversation with Miloš I asked him whether he was satisfied with the results and he responded that 'The reason I didn't expect the post-CCT [corrected chronologies] results to be better than they were is because the method is not perfect and actually any improvement at all is a good sign.' In fact, Miloš suspected that CCT could be 'over-correcting' the chronologies and removing part of the climatic signal. Miloš was planning to collaborate further with Dan to resolve this issue but 'for the purpose of cleaning the chronologies from disturbance', Miloš concluded, 'CCT is good enough for me now'.

Miloš first presented the results with CCT on the Scottish data at the international conference in Melbourne in January 2014, for which he received a prize from the scientific committee. A couple of conference attendees I talked to praised the fact that the CCT method 'bridged the gap' between dendroecology and dendroclimatology, and brought together the perspectives from these two subfields of dendrochronology. Understandably, Rob was very pleased with Miloš's prize and told me that he considered the CCT methodology to be 'the most innovative' element of Miloš's PhD thesis and speculated that the CCT could become a potentially revolutionary method as 'for a very long time, we assumed that our chronologies were disturbance-free, but this methodology might reveal that we've been wrong all along'. Rob imagined a situation in which all published tree-ring chronologies could be re-analysed with CCT and encouraged Miloš to start writing a paper about this method with Dan (eventually published as Rydval et al., 2016), in which Rob was included as a co-author (Box 6).

Box 6

How does the style of supervision affect the content of an epistemography?

'You have never co-authored a paper with your supervisors?' Miloš asked me in shock. This question made me think, for the first time, about our different supervision arrangements. As the sociologist Michael Buroway (2005) suggests, the production of dissertations in the natural sciences resembles the 'apprenticeship' model in which students work on and elaborate the research of their supervisor and, hence, jointly authored articles are the norm. In the social sciences, on the other hand, the dissertation labour process can be characterised as the 'independent scholar' model in which students obtain their own funding, choose their own research topic (often biographically related) and typically produce solo-authored publications. As Buroway (2005: 44) explains, 'Independent scholars take themselves as point of departure, whereas the apprentice looks to the master to initiate and then direct his or her research … In the former the focus is first on producing the product and second on producing the producer, whereas in the latter the order is reversed, training the sociologist comes first.'

The present epistemography has resulted from the two distinct styles or cultures of supervision broadly existing in the natural and the social sciences. As a social scientist, I was trained within the independent scholar model, but as a social scientist studying natural scientists I also underwent some training in the apprenticeship model. Because I worked as a 'voluntary technician' for Rob and Miloš, my research effectively became an extension of the Scottish Pine Project. So much so that Rob included me in the laboratory website as one of his students, and nowadays I appear as a defended post-graduate student (see picture) and my project became part of Rob's 'outreach activities' portfolio.

Throughout my research, I envied certain aspects of the apprenticeship model, as I saw it practised by Rob and Miloš. I was envious of the fact that Rob and Miloš saw each other every week, and that Rob was always available to Miloš for discussing his research (in contrast, I saw my supervisors roughly for an hour each month). I also admired the fact that Miloš's project was so evidently related to a bigger research question, and that the scholarly tradition to which he expected to contribute was so familiar to him, as he got to meet people like Malcolm Hughes and Edward Cook (instead, I had a virtual dialogue, through texts, with an invisible community of scholars, many of whom were dead or unknown to me). My unease with the independent scholar model, as I experienced it, was that I had fewer opportunities than Miloš for learning from my supervisors and reputed scholars in my field about the 'crafty' aspects of doing research and theorising (I always wished there was a 'fieldweek' in STS, which was equivalent to the dendrochronological fieldweeks described in Box 5!).

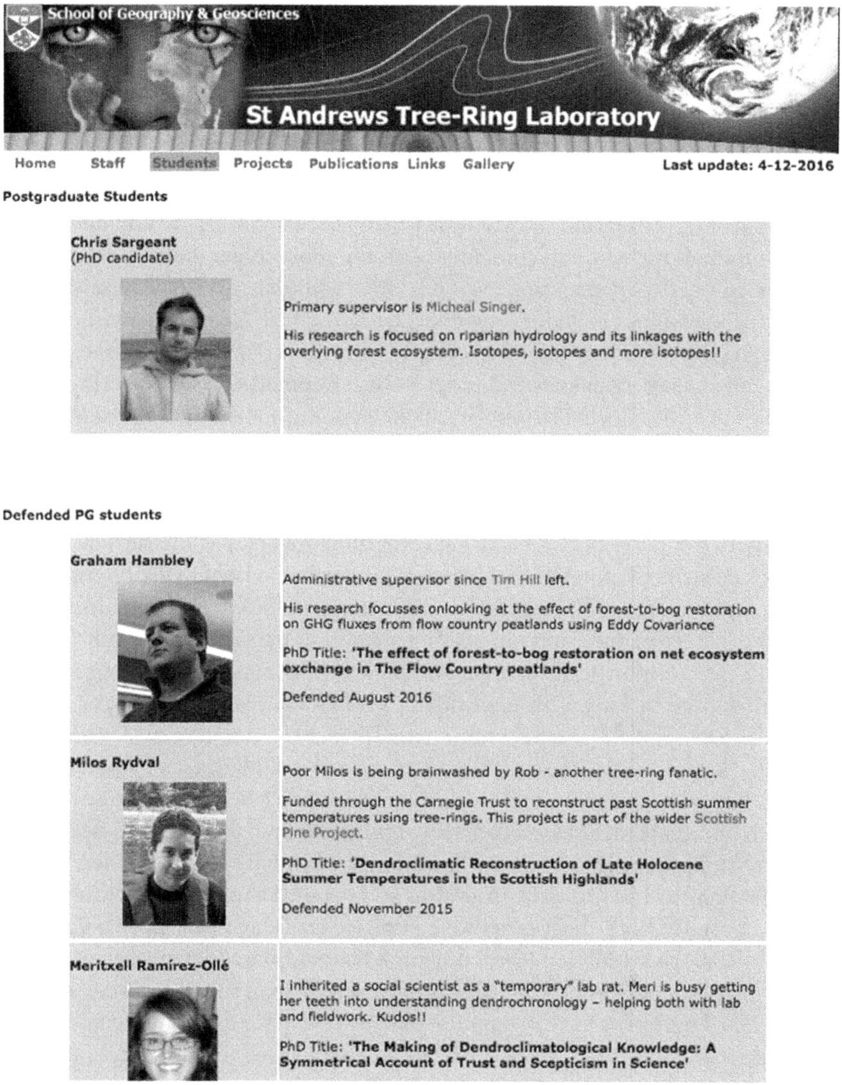

3.3 The author became included in the Scottish Pine Project as a student.

Becoming an independent scholar through the conduct of this epistemography did not mean learning in isolation or in a social vacuum. Mentoring always takes place in the broader institutional and political context of a country, university and department, which define the duties and responsibilities of the mentors and the mentee. In my case, I conducted my research at the University of Edinburgh in the period 2009–2015 with a scholarship from the UK's largest public funding body of social research (the Economic and Social Research Council) to which European Union citizens like myself were, at the time, eligible to apply (this might change after the 2016 'Brexit' referendum and subsequent negotiations) and that only covered tuition fees. Consequently, I had to work as a teaching assistant throughout my PhD, which meant, among other things, that I chose to conduct an epistemography of a scientific community located near where I lived and worked (Edinburgh). Even though my research coincided with the most recent global economic crisis, in my department there was a relative abundance of state-sponsored PhD scholarships (including my own) from the UK, Colombia and South Korean governments. We had to comply with strict requirements related to fees, curriculum, deadlines and progression/examination stages determined by the university and the respective national funding bodies. In turn, academic staff in my department absorbed this considerable mass of PhD students more or less evenly; my supervisors mentored between two and three students, which meant that they had limited supervision time.

My two supervisors had equal and yet different roles in the supervision of this epistemography, which, I believe, were mainly related to their training. One supervisor (a male professor and a historian of science) always insisted on the 'bigger picture' and on creating an overall narrative throughout my thesis, which influenced the final structure of this book about the process of knowledge-making. The other supervisor (a female lecturer, a biochemist turned STS scholar) insisted on the 'smaller picture' and gave me extensive feedback on specific aspects of the data, often resulting from comparisons with her own past experience as a biochemist and her current research, which forced me to develop a coherent argument within each chapter. Neither of my supervisors had any prior knowledge of climate science or dendroclimatology so I had to write in as clear and detailed a way as possible about my case study. Both supervisors were knowledgeable and participants in the local sociological tradition (the 'Edinburgh School'; Mazanderani et al., 2018) in which I was also being trained, which meant that our monthly discussions developed under the tacit assumption that my research was part of a common theoretical lineage, and, as such, my research would seek to improve it rather than to reject it (see Ramírez-i-Ollé, 2019b).

Having identified and partially eliminated the effect of disturbance on the tree-ring data, Rob and Miloš still faced the difficulty of choosing a standardisation curve to remove the ageing trend. Rob had reached the conclusion, after years of research, that 'there is no right or wrong way to detrend the data.' In all his previous work, Rob had used more than one standardisation curve to detrend chronologies. Rob explained, 'One colleague used to make fun of me that I could not decide which was the best version, but I always argued that they all had strengths and weaknesses.'

Rob proposed Miloš apply a relatively standard method in dendroclimatology called the 'ensemble approach' to resolve the indeterminacy of choosing one standardisation curve. Essentially, the ensemble approach consisted of generating variations of standardised chronologies with different standardisation curves (what Rob called 'flavours') and evaluating their adequacy in terms of their similarity (judged with correlation coefficients and other statistics) to temperature data. Miloš eventually experimented with four standardisation curves: two deterministic curves (the 'classic negative exponential' and 'Hugershoff' curve) and two non-deterministic curves ('Regional Curve Standardisation' and 'Signal Free').

The fact that the chronologies generated with the most recent standardisation method ('Signal Free') offered better results posed a problem for Rob and Miloš. As Rob explained it, 'The mantra is that Signal Free is a better method, but many people, including myself, don't fully understand how it works.' Signal Free was developed in the late 2000s by the dendroclimatologist Dr Thomas Melvin as part of his doctoral research, and it had the reputation of being a highly intricate method. In a lecture that Cook gave about Signal Free during the fieldweek I attended in Tasmania, he introduced Melvin's work as 'one of the most original and intriguing PhD dissertations ever done in the history of dendroclimatology'. Because of the apparent unintelligibility of Signal Free, Cook had collaborated with Melvin to develop a more accessible version, which became the version that Rob and Miloš used. In a conversation, Rob and Miloš discussed their concerns about obtaining better results with Signal Free because they could not defend such results to potential critical outsiders of dendroclimatology, whom they referred to as 'sceptics':

> Miloš [talking to Rob and pointing to the computer screen]: You see, these results are really interesting because, for some reason, with standard negative detrending there is not much of an improvement in the chronologies, but if you do it with Signal Free there is an improvement after cleaning them with CCT. But if you use the raw data [without CCT correction] and then apply Signal Free, then the results are much worse. I am not quite sure why and how to interpret these results …
>
> Rob: Well, I don't quite know Signal Free either. I've just toyed with it.
>
> Miloš: So, you know, I am a little cautious about these results.
>
> Rob: Yeah, yeah. No one really knows how it [Signal Free] works. I mean, people are black boxing it …
>
> Miloš: Well, I guess we will need to experiment with it then?
>
> Rob: Yes, I know how it works conceptually, but I don't think I truly understand how it works. [Three seconds of silence]. Yes, Signal Free is certainly going to be an issue. We can make subjective and objective decisions with regards to standardisation; I am not too concerned about this. But we've just got to rationalise every step.
>
> Miloš: Yes, yes [nodding].
>
> Rob: I am just thinking about the sceptics. We are actually in an interesting position because I know that Montford is very interested to see what comes out of Scotland. So we have to be clear about everything we do. I have actually agreed to write a blog post, but he will also keep an eye on our papers.

Miloš: Oh, yes [nervous laughing].
Meritxell: No pressure!
Miloš: Yeah [laugh], sure, no pressure at all!

As the dialogue above illustrates, Rob and Miloš's main concern was to understand Signal Free better and to make the various methods they had used to standardise the data publicly justifiable. Rob complained about the fact that colleagues, and even himself, were 'black boxing' Signal Free and regarded this attitude as problematic because it opened the door to criticism from 'sceptics' like 'Montford', as mentioned explicitly by Rob. Andrew Montford, who conducted a private investigation about the hacked climate science emails and published a book about it (Montford, 2010), is the owner of the 'Bishop Hill' blog and had met Rob a few times at the University of St Andrews, as he lived near Rob's workplace. Rob valued positively his interactions with Montford and some of his blog readers because, as Rob explained to me, 'these individuals have some faith issues with regards to the science but they are generally willing to learn, and their minds are not closed'. In this way, on Rob's invitation, Montford had once participated in a university panel about climate change and had agreed to publish a blog entry from Rob's undergraduate students where they had to respond to questions and criticism formulated by blog readers. The purpose of all these interactions was as Rob explained to me, 'to get students thinking about how to deal with sceptics'; much in the same way, Rob had encouraged Miloš in the conversation above to consider how 'sceptics' would react to their standardisation choices, and try to pre-empt their criticisms as much as possible.

Conclusion

This chapter has detailed the process of analysing tree-ring data and, specifically, discerning which growth trends in tree-ring chronologies are (ir)relevant for reconstructing past climate in Scotland. Standardisation, I suggest, is essentially an act of classification. As we have seen in this chapter, standardisation has involved the ordering of data using the categories of 'noise' and 'signal', which seem to be a form of classification commonly used by natural and social scientists alike (Silver, 2012; Collins, 2004b: 277; Knorr Cetina, 1999: 46–78; Smith, 1997: 11).

To classify between unwanted, polluting, dirty 'noise' and desirable, clean, meaningful 'signal' is not only a scientific practice but also a human constant. The 'signal/noise' boundary is akin to the 'cleanliness/dirt' distinction that the anthropologist Mary Douglas argued to be constitutive of mundane social life. For Douglas, the human aversion to dirt is universal because we are all intolerant of disorder, but the exact definition of dirt is variable according to different historical and social contexts: 'There is no such thing as absolute dirt: it exists in the eye of the beholder … In chasing dirt, in papering, decorating, tidying we are not governed by anxiety to escape disease, but are positively re-ordering our environment, making it conform to an idea. There is nothing fearful or unreasoning in our dirt-avoidance: it is a creative movement, an attempt to relate form to function, to make unity of experience' (Douglas, [1966] 2001: 2). Similarly, the meaning of 'noise' and 'signal' with regards to tree-ring data is contextual and depends upon the researcher's goals. As Cook expressed: 'one researcher's "signal" would frequently be another researcher's "noise"'. While disturbance was a 'leftover' for Rob, for the dendroarchaeologist Anne and the dendroecologist Dan, disturbance was the desired clean signal. Overall, Cook's conceptual model of tree growth is the main classificatory system that dendroclimatologists like Rob and Miloš employ to understand and to impose a certain order on the tree-ring data that they seek to interpret.

If we were to look at standardisation as classificatory work, as I suggest, we would be able to discern more clearly the efforts and social dynamics sustaining such work. Once they become accepted, classifications often become reified and the motivations and procedures behind their creation become invisible. For this reason, as Geoffrey C. Bowker and Susan Leigh Star (1999: 34) propose, the investigation of classification systems (they call them 'infrastructures') should follow the method of 'infrastructural inversion', which they define as 'a struggle against the tendency of infrastructure to disappear (except when breaking down)' and it 'means recognizing the depths of interdependence of technical networks and standards, on the one hand, and the real work of politics and knowledge production on the other'. Bowker and Leigh Star (1999: 5) exemplify this method: 'Much as a city planner or urban historian would leaf back through highway permits and zoning decisions to tell a city's story, we delve the dusty archives of classification design to understand better how wide-scale classification decisions have been made.'

In foregrounding the socio-material practices involved in sorting out and eliminating noisy trends from the Scottish dataset, the importance of familiarity, personal reputation and access to relatively private communication spaces in academia such as workshops and laboratory visits becomes evident (Collins, 2001, pointed out long ago that workshops and laboratory visits are a key mechanism for establishing bonds of trust between scientists). As shown in this chapter,, the standardisation work that Rob and Miloš conducted was based on the trust that generations of dendroclimatologists have placed on Cook himself and his conceptual model (hence why the model has become included as a 'principle' in the most recent textbook; see Speer, 2010: 17). Rob and Miloš specifically benefited from Cook's brokering role in order to eliminate the disturbance and the ageing trends: Cook recommended CCT to Rob in a workshop and, as a director of the tree-ring lab at Lamont, Cook indirectly facilitated Miloš's meeting with Dan and Kevin. Likewise, Cook also created a more accessible version of Signal Free, which Rob, Miloš and many others currently use.

This chapter has also shown the personal and professional interests sustaining the work of standardisation that Rob and Miloš carried out in Scotland. Their decision to employ the CCT method and the ensemble approach could be likened to an 'investment', intended to yield recognition from peers and outsiders (or, using Pierre Bourdieu's terms, to yield 'social and cultural capital'; 1986) which, in turn, could bring about more 'economic capital' (i.e. publications, grants, students, funds and a bigger laboratory) not only to themselves but also to the broader community of dendroclimatology. To think of scientific work as a series of interlinked investment strategies (Latour and Woolgar, [1979] 1986: 187 talked about 'cycles of credit') is appropriate if one is careful not to think of the individual scientist as an independent goal-seeker or investor. A 'good' scientist, as valued by his or her peers, is someone who takes into consideration and seeks to address common goals and challenges. In this way, Rob and Miloš's interest in employing the CCT method was not restricted to the elimination of disturbance from the Scottish dataset, but, as Rob made explicit, he hoped that CCT could revolutionise the entire field, as he imagined a situation in which colleagues could also benefit from discovering that their chronologies contained disturbance. Similarly, Rob's suggestion to use the ensemble approach sought to resolve an important problem faced by all dendroclimatologists.

If we look at standardisation as a form of classification, we will also be able to unveil the ways in which the work of eliminating noise disturbance from the Scottish data was used for organising people and acting upon the world. Classifications are not only used for describing the world, but also for arranging it. For instance, we wake up at 7am and not 8pm and we organise our daily schedule accordingly; we think it is best to eat 'healthy' instead of 'junk' food and our lifestyles are based upon this distinction; and the 'British' are, of course, taken to be different from the 'Lithuanians' in passport control. Such mundane classifications have profound political and moral consequences, which only become visible to us when we transgress them. Bowker and Leigh Star (1999: 3) again usefully illustrate the coercive power of classifications: 'Try the simple experiment of ignoring your gender classification and use instead whichever toilets are the nearest; try to locate a library book shelved under the wrong Library of Congress catalogue number; stand in the immigration queue at a busy foreign airport without the right passport or arrive without the transformer and the adaptor that translates between electrical standards. The material force of categories appears always and instantly.'

The work of standardisation described in this chapter effectively valorised the point of view of dendroclimatologists and silenced, to a certain extent, the interests of other tree-ring data users. Tree-ring data can be used for different purposes and each application has developed into subfields of dendrochronology: dendroclimatology (the study of past climates); dendroarchaeology (the study of past societies); dendroecology (the study of woodland ecosystems); dendromorphology (the study of land movements); dendrochemistry (the study of pollution among other topics). In the two dendrochronology conferences I attended in 2014, talks were arranged into streams that corresponded to each of the subfields, and plenary sessions always included a speaker from each specialised area. Despite all attempts at equitability, a few attendants to whom I talked at these conferences expressed their resentment of what they perceived as the predominant position of dendroclimatology. One dendroarchaeologist said that he was 'bored' of going to conferences and just listening to colleagues talking about 'low and high frequency', which are terms that only dendroclimatologists would use. Another dendroarchaeologist also complained that 'now it seems that chronologies are worth nothing unless you use them for climate reconstructions'. A junior dendroecologist also complained that ordinary people associate the use of trees with climate reconstructions, and insisted that 'there is much more to learn from trees than just climate'.

Precisely, I interpret the perceived originality and success of Miloš's work on the CCT method – as seen by the fact that the scientific committee at the conference awarded him a prize – as an indication that this method was seen as an opportunity to combine perspectives in the analysis of tree-ring data and to equalise relationships between dendroecologists and dendroclimatologists. In short: the CCT method, as a socio-material technology of classification, was seen as the means for re-ordering human interactions among specialists, or, as one dendroecologist put it, 'to bridge the gap' between dendroclimatologists and dendroecologists.

With this discussion of the social nature and consequences of the classificatory work of standardisation I would like to question the image of climate science as an impersonal, disinterested and contemplative endeavour. In the weeks and months after the hacking of climate science emails in November 2009, many scientists referred to the social factors and dynamics present in science as corrupting factors; as one physicist put it: professional rivalries, limited time and passionately held beliefs are 'things [that] can *prevent us* from finding out what works' (Butterworth, 2010, emphasis added). The existence of political and professional interests was presented as a criterion for identifying illegitimate criticism of climate science (Ramírez-i-Ollé, 2015a). The email hacking itself was interpreted by one prominent climate scientist as the result of a 'campaign run by economic, political, and ideological interests opposed to many proposed policies that might deal meaningfully with the threat of climate disruption' (Sommerville, in Yale Forum on Climate Change & The Media, 2010). 'Good' climate science was defined as being developed autonomously from society: the climate scientist Michael E. Mann published an article defending that 'Science must be unpolluted by politics' (Mann, 2009), and another climate modeller argued that 'For me, good science means generating knowledge through a superior method, the scientific method. The merits of a scientifically constructed result do not depend on its utility for any politician's agenda … As a scientist, I strive for independence from vested interests. I am in the pocket of neither Exxon nor Greenpeace' (Von Storch, 2009).

The common prejudice that regards social processes and influences as *only* distortions to the production of knowledge (this preconceived opinion has been called the 'sociology of error'; see Kamwendo, 2017; Bloor, [1976] 1991: 13) is false insofar as it is one-sided and does not consider the beneficial and necessary influences of the scientific community and the wider society on scientists and their work. How could Rob and Miloš have successfully eliminated noise from the Scottish dataset and contributed to the development of CCT if not by trusting Cook's conceptual model and using their relations with him to facilitate a meeting with Kevin and Dan at Lamont? How could a scientist more generally dispense with the recourse to familiarity without ruining their working relations and becoming an isolated and unproductive individual? After months and years spent refining theories, experiments and techniques like CCT, why do we expect scientists like Dan, Miloš and Rob to be dispassionate about their creations and not use them to address personal, professional and social problems? If one ignores the reputation and past productivity of an individual or research group, on what efficient grounds can someone evaluate their claims? In a world of overload of information, how are scientists expected to limit irrelevant data and discriminate on which aspects are worth investigating if not by relying on trustworthy advice? Why would we deprive scientists of the very useful qualities of being part of a community of practitioners that gives them access to communication channels that filter reliable judgement and advice? Why do we not simply accept that scientists, like any other professional group, have a collective interest (note the etymological derivation of 'interest' from 'inter-esse', which means 'what lies in between actors and their goals; Latour, 1987: 108) in actively defending their credibility as shown in the recent global Marches for Science (Niiler, 2017)? Overall, why do we not regard the social/emotional and the rational/cognitive dimensions of science as mutual constituents rather than in opposition?

I think it would be counter-productive for scientists to ask them to be unconcerned by whatever happens to their fellow citizens and peers. As Daniel Sarewitz (2016) argues, the ideal that 'scientific progress on a broad front results from the free play of free intellects, working on subjects of their own choice, in the manner dictated by their curiosity for exploration of the unknown is a bald-faced but beautiful lie upon which rests the political and cultural power of science'. Sarewitz continues, 'Scientific knowledge advances most rapidly, and is of most value to society, not when its course is determined by the "free play of free intellects" but when it is steered to solve problems.' In other words, successful science is not curiosity-driven but goal- and problem-solving-driven. In relation to this chapter, for instance, Rob and Miloš sought to address two goals and problems. First, they had the immediate goal of standardising tree-ring data. Secondly, they had the longer-term goal of learning about regional climate change in Scotland. In turn, Rob and Miloš's aim of reconstructing the Scottish climate is part of a broader epistemico-political-social goal: making global environmental science and politics (Beck et al., 2016). As the historian of science Jon Agar (2012: 397) has shown, the monitoring and maintenance of global order and environmental stability was (and perhaps still is) one of the key social problems of the twentieth century, to which many contemporary sciences, including dendroclimatology, are orientated. Social and political concerns are, therefore, one of the multiple causes of formation of scientific knowledge about climate change, and without them climate science as we know it would not exist.

4

Reconstruction

After having sampled trees, dated tree-rings and cleaned tree-ring data, Rob and Miloš were finally at the stage where they could discover how the climate in Scotland was before the early 1700s, when methodical thermometer-based records began. Dendroclimatologists refer to the work of re-enacting past climate from tree-ring data as 'reconstructing' and the resulting graph a 'climate reconstruction'. As told in previous chapters, Rob and Miloš built upon Hughes's pioneering reconstruction of past temperatures in Scotland, which reached the eighteenth century (Figure 4.1). They hoped to use the 800-year-long chronology from the Cairngorms (Chapter 2) to extend Hughes's temperature reconstruction back to the thirteenth century. On the website of the Scottish Pine Project, Rob wrote, in reference to Figure 4.1, 'The original Hughes reconstruction is shown below showing the excellent calibration potential of this species [Scots pine trees] in the Scottish Highlands. Although some success has already been made in finding older living sites, the truly exciting work will be related to extension of the living material with either historical or sub-fossil material' (Scottish Pine Project, 2019).

The work of reconstructing climate, as described in all dendroclimatology textbooks and in Rob's classes, involves the stages of 'calibration' and 'verification'. The stage of calibration consists of using the *first half* of the available local meteorological dataset (precipitation or temperature) and to compare it with or 'calibrate' it to tree-ring data. The stage of 'verification' involves comparing the tree-ring based climate reconstruction against *the other half* of local meteorological data withheld from calibration. In its simplest form, both stages are computerised using a statistical technique called 'linear regression analysis'. The computer creates a 'response function' or a mathematical equation that models how the tree 'responds' to temperature/precipitation data during the calibration period. Using the technique of linear regression analysis, the computer reconstructs past climates by inverting the calibration equation and using tree-ring data as the predictor and instrumental data as the predictand. The resulting equation is called the 'transfer function' as the tree-ring data are 'transferred' into estimates of past temperature or precipitation. Dendroclimatologists use correlation coefficients and other statistics to verify the 'skill' or the ability of the climate reconstruction to resemble the instrumental data used for the verification period.

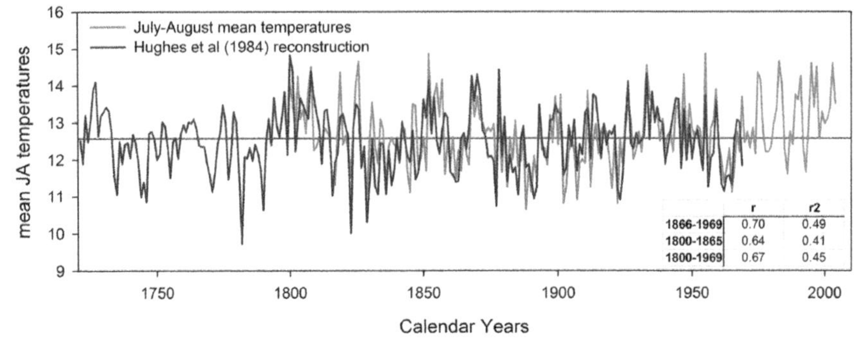

4.1 Rob and Miloš's work built upon an existing reconstruction of the Scottish temperature.

The work of reconstructing past climates from tree-ring data is based on the supposition that meteorological records are the 'golden standard' against which to calibrate and validate a climate reconstruction. While the possibility of measuring temperature accurately was a contested scientific issue throughout the eighteenth and nineteenth centuries (Chang, 2004), nowadays the reliability of meteorological data for measuring twentieth-century global warming is largely undisputed (experts, of course, still investigate problems with global temperature datasets; see Jones, 2016). In paleoclimatology, thermometers are regarded as the best source for generating *direct* evidence of past and present climate in the form of so-called 'instrumental data'; in contrast, trees, ice cores and marine sediments, among others, are described as 'proxies', 'natural archives' or 'paleothermometers' from which to generate *indirect* evidence of climate (Bradley, 1999).

The second supposition necessary for generating tree-ring-based climate reconstructions is that the present-day relation between climate and tree growth will extend backwards beyond meteorological data. This assumption, shared with paleoclimatologists and other natural scientists studying past phenomena, receives the name of the 'principle of uniformitarianism'. The definition and authorship of this principle have long been a source of dispute in geology (Shea, 1982; Gould 1965). Dendroclimatology textbooks, however, offer a simplified definition of uniformitarianism as 'the present is the key to the past', a sentence attributed to the eighteenth-century Scottish geologist James Hutton (Speer, 2010: 10). The dendroclimatologist Harold Fritts (1976: 14–15) clarifies that 'Applied to dendrochronology, the uniformitarian principle implies that the physical and biological processes which link today's environment with today's variations in tree growth must have been in operation in the past.' He insists that uniformitarianism does not imply that the paleoclimate was the same as the climate of the present. However, 'it does imply that the same *kinds* of limiting [environmental] conditions affected the same *kinds* of [physical and biological tree growth] processes in the same ways in the past as in the present'.

Over the years, dendroclimatologists have realised that the uniformitarian principle does not always hold true. The constitution of the 'divergence problem' as a research topic in dendroclimatology exemplifies this collective acknowledgement. The divergence problem refers to the observation that, in some sites in the Northern Hemisphere, tree-ring data and temperature trends appear to have diverged since the 1960s. While temperature has been recorded by thermometers as steadily rising, some tree-ring data show a declining trend. Consequently, trees might not seem to record recent warmth. Dendroclimatologists – Rob included, as his still most cited paper is a review of this research problem (D'Arrigo et al., 2008) – have proposed numerous theories to explain the phenomenon of divergence. Divergence is still a topic of concern and study for dendroclimatologists and, as Rob stated in a plenary talk at a conference in Melbourne in January 2014, 'I think we do know much more now than we did in 2008.'

The identification of a divergent trend between temperature and certain tree-ring datasets since the 1950s has led some commentators outside the dendroclimatology community to question uniformitarianism and dendroclimatology altogether. Steven McIntyre, the owner of the blog 'Climate Audit', has focused, among other topics, on the 'divergence problem' (https://climateaudit.org/category/proxies/divergence/). Other bloggers criticise the assumption of linearity, as '[dendroclimatological] studies universally assume a linear relationship between climate and ring response and then invert this relationship in order to predict past climate states. The point is that this assumption/practice is *not justified*, either empirically or theoretically; it constitutes a serious conceptual mistake in tree ring analysis' (Bouldin, 2012).

While dendroclimatologists do not regard the phenomenon of divergence as a complete refutation of uniformitarianism and dendroclimatology, they admit that it raises some doubts about the development of tree-ring-based climate reconstructions. The author of the most recent dendrochronology textbook acknowledges that the climate has indeed changed over time as a result of unusually elevated levels of carbon dioxide and other greenhouse gases in the atmosphere, and that this change might alter the way trees respond to climate change (and hence violate the principle of uniformitarianism), but he defends the uniformitarianism principle as a 'productive starting point'. In his own words, 'We know that our assumptions that present processes have not changed through time is not always correct, but uniformitarianism is a *productive starting point* in the analysis of past climates and environmental variability' (Speer, 2010: 11: emphasis added). In their review article, Rob and colleagues (D'Arrigo et al., 2008: 296) restrict the problem of divergence in ring-width and density-based chronologies from certain Northern forests, and concede, 'The principal difficulty is that the divergence disallows the direct calibration of tree growth indices with instrumental temperature data over recent decades (the period of greatest warmth over the last 150 years), impeding the use of such data in climatic reconstructions. Consequently, when such data are included, a bias is imparted during the calibration period in the generation of the regression coefficients.'

Dendroclimatologists hope that the development of new tree-ring data will allow them to determine with more certainty the reliability of their reconstructions. For this reason, Rob and colleagues (D'Arrigo et al., 2008: 302) finished their review article about the divergence problem by stating, 'As existing records are updated and new ones developed, we will improve our ability to make more defined, direct evaluations of the climate of recent decades relative to the past.' Similarly, all the dendroclimatologists I saw presenting their reconstructions at conferences ended their talks by saying something like 'more data are needed to draw more definitive conclusions'. For this reason, Miloš also kept generating new tree-ring chronologies until the very last few months of his PhD. The reason why Miloš submitted his thesis later than he expected was that he waited for Rob to crossdate the 'floating' subfossil chronologies from the Cairngorms (Chapter 2). Rob and Miloš saw the constant flow of 'more data coming in' as a requirement for building up a cleaner climate signal that offered better calibration and verification statistics. The better the statistics, the more certain they could be that the reconstructed temperature values beyond the calibration and verification periods were accurate.

The epistemological conundrum that Rob and Miloš faced at this stage of the production of dendroclimatic knowledge was, as they assembled more data, to decide whether the finite and limited evidence they had about the present relationship between climate and tree growth could be extrapolated into the past beyond the calibration and verification periods, and ultimately to convince themselves, colleagues and the broader public that their temperature reconstruction was a truthful representation of past climates in Scotland. Rob and Miloš sought to reduce the uncertainty inherently involved in reconstructing past climates, I suggest, using a double strategy: 'trained variation and natural selection' and 'comparability work'.

Trained variation and natural selection

I characterised Rob and Miloš's initial work of reconstruction as 'trained variation and natural selection' after reading a sociological study of the work of molecular biologists (Box 7). The author of that study, the sociologist Karin Knorr Cetina (1999: 91), explains: 'If there is a general strategy molecular biologists adopt in the face of open problems, it is a strategy of blind variation combined with a reliance on natural selection. They vary the procedure that produced the problem, and let something like its fitness – its success in yielding effective results – decide the fate of the experimental reaction.' Later on Knorr Cetina (1999: 109) clarifies that rather than deploying 'blind variation', these molecular biologists draw upon their trained expertise to generate variations that they think will be more likely to be selected by nature: 'Variation in molecular biology, however, is by no means as sightless and undiscerning as the random genetic mutations from which the term *blind variation* is borrowed. For example, the experienced body of the scientist, when it operates, naturally brings its experience to bear on the variations it concocts for selection by success. The retries scientists perform are never just any odd random alterations. Instead, they are based on what a scientist "senses" to be a promising strategy in a problem case.' Similarly, I show below that Rob and Miloš produced multiple versions of the reconstruction with the aim of finding the one that had the best 'fit' and resembled more closely 'nature' (i.e. instrumental temperature data). However, rather than selecting themselves the 'best' reconstruction, Rob and Miloš delegated this decision to proxy 'nature' (i.e. the climatic signal assumed to exist within tree-ring chronologies).

Box 7

Creating epistemographic knowledge by way of analogy and induction

I read epistemographies for inspiration throughout my research. A book that I found particularly inspiring was Karin Knorr Cetina's *Epistemic Cultures: How the Sciences Make Knowledge*, which is the first ever comparative study of two scientific cultures (molecular biology and high energy physics) and their organisational strategies for making knowledge. I came to use this book as a 'template', following the author's own advice. In a dialogue with a fictional reader, Knorr Cetina suggests (1999: 252), 'I would use the results – not by generalizing them, but by using the patterns I illustrate as templates against which to explore the distinctive aspects of other expert domains, and as pointers to possible dimensions in other areas.' For instance, the notion of 'trained variation and natural selection' that I took from Knorr Cetina 'can be used as sensitizing concepts in any other study of epistemic cultures to determine how these cultures are configured, what similarities might exist, and how to account for them' (1999: 252).

I employed epistemographic knowledge as 'sensitizing concepts' or interpretative devices that helped me to make sense of both the specificities and similarities of my epistemic object of study in comparison with other objects. The sociologist Herbert Blumer (1954: 7) defined the function of 'sensitizing concepts' as 'it gives the user a general sense of reference and guidance in approaching empirical instances'. As I understand it, sensitising concepts that are used as a starting point in one research context can then be developed into more or less 'definitive concepts' or theories about similar research objects. Blumer (1954: 7: emphasis added) defined 'definitive concepts' as 'refer[ring] precisely to what is *common* to a class of objects'. One of Blumer's students, the sociologist Howard Becker, further developed the idea that theorising essentially depends upon exploring the differences and the commonalities between social phenomena. Becker (1998: 126–127) starts by recounting the advice that one of his colleagues gave to students ('Tell me what you've found out, but without using any of the identifying characteristics of the actual case') and employing this advice to theorise about two social activities, banking and baseball:

> I would have to choose words more general than the specifics of my case, but not so general as to lose the specificity of what I found [...] That's how I made the move from the fact that banking executives steal to the statement I made about the clarity or ambiguity of an action's criminality. I restated

> the assertion that 'the executives of savings and loan associations sometimes steal money by manipulating banking regulations whose complexity makes it difficult for prosecutors to decide whether what they indisputably did is a 'crime' without using any of the specifics. I didn't say 'executives' or 'savings and loans' or any of the other specifics. I said what class each of those belonged to and so ended up talking about the ambiguity of an action's criminality, a dimension that could be useful in the study of any criminal activity. And I could take another step and talk about something less specific than criminal law – rules in general – and that would let me introduce such interesting cases as whether the ball the pitcher throws is a 'ball' or a 'strike,' the rules for deciding that being as ambiguous as any in the criminal law. You could argue that, after all, baseball and banking don't have much in common … Both the similarity and the difference give us general categories to think about and use in our analyses. The similarity says, by way of generalizing, 'Every set of rules is clear to some degree and ambiguous to another degree.' The difference says, by way of a different kind of generalization, 'Within the organizations (like baseball and banking) in which rules are made and enforced, there are other things going on, such that those rules will vary along a dimension running from clarity to ambiguity.' Making such comparisons reveals further complexities in the creation and application of rules, complexities that can be attended to in future research. The immediate consequence of that result is that every study can make a theoretical contribution, by contributing something new that needs to be thought about as a dimension of that class of phenomena.

Like Becker above, when I compared Rob and Miloš's reconstruction work to the molecular biology work studied by Knorr Cetina I concluded that, despite their significant differences, they both represented the epistemic strategy of 'trained variation and natural selection'. I also developed the other concepts and epistemic strategies presented in this book (i.e. the chapter sub-headings) by way of analogy and induction from other empirical studies and scholarly ideas (as indicated in the in-text references). These concepts are what the sociologist Robert K. Merton (1968: 39) referred to as 'middle-range theories', which seek to explain delimitated aspects of the data rather than offering catch-all explanations of social phenomena (I have offered elsewhere a general theory of epistemic-making, which I have called the 'theory of the externalisation of trust relations'; Ramírez-i-Ollé, 2018).

The strategy of natural selection and trained variation used by Rob and Miloš initially consisted in choosing the appropriate climate variable to reconstruct, either rainfall or temperature. Dendroclimatologists only reconstruct one climatic variable for a region because they believe that only the most limiting, climatic variable affecting the growth of trees in a certain area can be reconstructed (this assumption is known as the 'principle of limiting factors'; Speer, 2010: 15). Dendroclimatologists know that some tree species growing in certain locations are not suitable for reconstructions precisely because they have a 'mixed climate signal' and it is difficult to know if their growth is limited by either temperature or precipitation. For instance, as late as the 1980s, dendroclimatologists thought it was impossible to carry out dendroclimatology in the United Kingdom and the Republic of Ireland because trees, particularly oaks, showed a mixed signal (Briffa, 1984). This vision changed when Malcolm Hughes published the temperature reconstruction for Scotland in 1984, demonstrating that Scots pine trees growing at high elevations are limited by temperature and could be used for dendroclimatological purposes (Hughes et al., 1984). As part of their attempt to build upon Hughes's work, Miloš and Rob also reconstructed the climate variable of temperature.

While choosing 'temperature' as the climate variable to reconstruct was not problematic, Rob and Miloš struggled to choose the 'target season' or specific monthly meteorological data against which to calibrate tree-ring data. Their aim was to select the season in which tree-ring growth was most clearly limited by temperature. Rob and Miloš referred to this selection as 'maximising' the climate signal. The difficulty that Rob and Miloš faced was that each tree-ring parameter was affected differently by temperature. To illustrate this problem in a conference presentation, Miloš created a graph that showed the different responses of each tree-ring parameter to temperature data (Figure 4.2: 'RW' refers to ring width; 'BI' is blue intensity; and 'MXD' is maximum latewood density). While ring-width data correlated more or less uniformly throughout the year, blue intensity and density correlations were distinctly higher in July and August. Miloš rationalised this result in terms of the different physiological basis of parameters: whereas ring width is based on the cell growth of trees that can be triggered by favourable conditions throughout the year, blue intensity and density data are an expression of cell thickness and lignin content, laid down at the end of the growing summer season. Miloš eventually chose 'July–August' as the target season based on Rob's

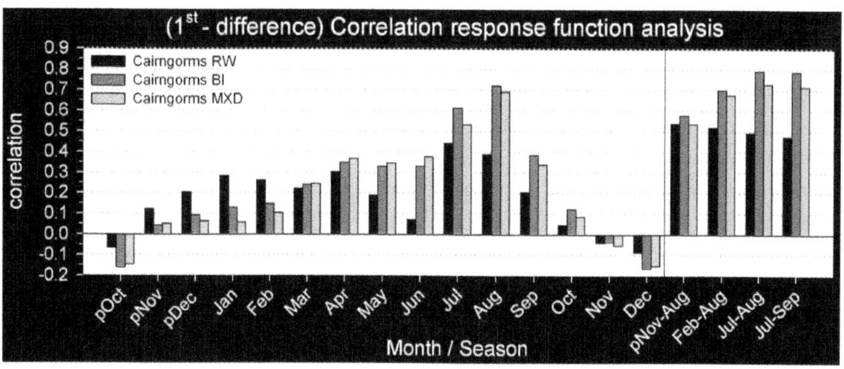

4.2 Dr Hughes's work was a useful precedent for choosing the season and climate variable.

recommendation: 'For the sake of coherence with Hughes's work I think it's best to choose the same months he used.'

The method of linear regression employed to reconstruct climate requires working with *averaged* series of meteorological and tree-ring data. With regards to tree-ring data, Rob and Miloš employed the 800-year-long chronology from the Cairngorms (Chapter 2). With regards to temperature or instrumental data, Miloš and Rob employed a 200+-year-long record of average monthly temperature data series from different weather stations in mainland Scotland. Rob and Miloš were very pleased to have at their disposal the second-longest temperature series in the UK (starting in 1800) because the longer the instrumental data, the more evidence they could obtain about the extrapolations outside the calibration and verification periods.

Rob and Miloš discovered a divergence in the early nineteenth century between the tree-ring data and the monthly temperature record from Scotland. As Rob knew one of the scientists in charge of curating the temperature dataset, Rob wrote him an email asking for any explanation he could offer about this divergence. The other scientist pointed out that, in one of his published papers, he had suggested the possibility that the temperature measurements from 1800 to 1866 might have been inflated because the weather stations of the time were not properly insulated from the sun. Rob responded, 'Your paper [Jones and Lister, 2004] is perfect as we actually have found a misfit between the TR [ring width] based reconstructions (too cold) and the instrumental data prior to about 1857 and your paper says that there are homogeneity problems prior to 1866. This is great as it was a real worry for us.' Rob reported to Miloš by email the discovery of the warming bias in the temperature series as 'great news for us' as he saw the anomaly in the temperature dataset as an indication that 'the divergent trends we get in the calibration period are not due to problems with the tree-ring data and validate the issues raised in the Jones and Lister [2004] paper'.

With regards to the averaged tree-ring data, Rob and Miloš faced the difficulty of choosing one standardised chronology. As we have seen in the previous chapter, rather than choosing one single standardisation method, Rob proposed to use the 'ensemble approach' in order to generate multiple standardised chronologies. In total, Miloš created seven variants of the Cairngorms chronology and they pondered whether to generate seven different reconstructions (from each chronology variant) or generate a single one. Eventually, Rob decided to generate a single reconstruction using a relatively new method that he called the 'combo approach'. Essentially, the combo approach consisted of creating an averaged reconstruction from a weighted combination of all the temperature reconstructions derived from the 'ensemble approach'. The procedure started by ranking the seven standardised chronologies in terms of their similarity to temperature data. Rob and Miloš evaluated this similarity with a statistic called 'root mean square error'. The chronology with the smallest square error was weighted most heavily in the average of all the reconstructions.

Effectively, the combo approach epitomises the strategy of 'natural selection', as the weighting procedure spared Rob and Miloš from making a decision about which reconstruction version to accept as more truthful. Instead, the decision on the 'best' representation of climate was left to 'nature' (the climate signal within the tree-ring chronology that would correlate more strongly against temperature data). The reviewer of the article where Rob first presented the combo approach also implicitly recognised this feature when they said: 'For this reviewer this is a new and important concept that encourages experimentation and thinking about a range of reconstruction techniques and not having to pick the "best" reconstruction.'

As Rob described it to me, the combo approach was his attempt to develop a more 'objective' method for choosing the final reconstruction. 'In my COMBO games', Rob explained in an email, 'I derive multiple versions of the reconstructions and an objective way (hopefully) to combine them.' Rob was critical of the approach used by his colleague and friend Dr Jan Esper, which he informally referred to as 'Esperism'. When I asked Rob to clarify his disagreement with Jan, Rob told me that, while Jan had also created multiple reconstructions from multiple standardisation methods in one of his papers (Esper et al., 2007), he had picked one reconstruction on the basis of his 'expert judgement'. Rob criticised this choice because 'There was no statistical reason to choose this over other versions as far as I can tell. Jan feels he knows a "best" option which I feel we cannot do' (see Phillips, 2019 for a recent study of the supposed tension between 'human judgement' vs 'statistics' as evaluation tools). Rob was openly critical about Jan's approach, and the first time I heard Rob voicing his criticism was in one of his talks at the conference in Melbourne:

> I want to talk about the strategy that I am slowly starting to put together about how to go about the choice of the best reconstruction. With every record, whether it is a single site record or a regional composite, we need to be very careful in our local calibration and screening. We need to come up with the most robust calibration and verification statistics for a particular region. We can easily identify divergence empirically. We look carefully at residual analysis and through stringent verification we can say how robust a particular series is. We must use the 'best' chronology variant for a particular region.
>
> There are several approaches to this. We can use what I call 'Esperism' or the 'Esper approach' and his terminology of 'expert judgement'. Jan would say 'this is my best record' or whatever … [giggles from the audience]. I would much rather go for the statistical approach. Maybe we can use some sort of fusion using regression-based methods where you can combine ring width and density. We can maybe weight all the different variants as a function of their r^2 (correlation coefficients) against the target season, even weakly correlated chronologies can be still included, but they will be weighted very weakly. There are a few groups that are playing around with the band-pass approach; you might want to use the high frequency of density or blue intensity and the low frequency from ring width. [3 seconds of silence] These are all valid methods.
>
> Whatever you choose, you want to come up with the best explained variance that can be rationalised and is well defendable. I always put my Steve McIntyre's cap on when I do any analysis: 'Can I defend this in a public venue?' 'Is this a good choice? Do we have good reasons for doing it?'

In his speech above, Rob mentioned the blogger Steve McIntyre, and, more specifically, Rob said he always put 'my Steve McIntyre's cap on' when evaluating the adequacy of the final reconstruction. In this way, even though Rob thought that the final choice of a reconstruction should be left to 'nature', he considered his expertise to be necessary to explain 'nature's choice' to sceptical audiences like Andrew Montford and Steve McIntyre.

Comparability work

In the final stage of their reconstruction work, Rob and Miloš looked for additional evidence of climate change in Scotland against which to compare and evaluate the 'best' temperature reconstruction they had obtained with the combo approach. I follow Willem Schinkel (2016) in describing these comparative practices – which he argues are characteristic of paleoclimatology more generally – as 'comparability work'. The purpose of the comparisons carried out by Rob and Miloš, as Rob described it to me, was: 'We are not trying to use the other datasets to validate ours. We are more interested in the similarities and differences, like if we were comparing thermometers: is your thermometer seeing the Little Ice Age here?'

At the conference in Melbourne in January 2014, Miloš first presented some of the evidence resulting from comparability work. This was also the time when Miloš first presented a 'very provisional' 600-year-long temperature reconstruction for Scotland. At that point, the Cairngorms reconstruction only reached back to 1450 AD (this was before Rob managed to crossdate the floating chronology, Chapter 2). The graph that Miloš showed included two reconstructions: one was the 'Abernethy/Rothiemurchus' that Miloš had generated from the subfossil samples and living trees from two areas of the Cairngorms (Abernethy and Rothiemurchus); the other reconstruction included data from all the sites in the Cairngorms *except* Abernethy and Rothiemurchus (Figure 4.3). The graph was accompanied by a table that quantified the similarity between reconstructions and against the temperature dataset with correlation coefficients (r^2) and an associated statistic (Durbin–Watson or DW).

By excluding the Abernethy and Rothiemurchus chronologies from the broader Cairngorms dataset, Miloš and Rob created what they called an 'independent' reconstruction. Miloš interpreted this comparison positively because the two reconstructions showed very similar statistics as they were both from the same area in the East of Scotland, where trees had supposedly been affected by the same environmental conditions. In the question and answer session following Miloš's presentation in Melbourne, Malcolm Hughes, the author of the first Scottish reconstruction, asked him a question. Rob had ironically warned Miloš of this possibility as 'I am sure that Malcolm will ask you about his baby.' As a result, Miloš was very excited to meet Hughes and to discuss his work with him in the conference (Box 8).

Box 8

What conferences have done for this epistemography

Like Miloš, I met experienced and young colleagues at conferences. From the start of my PhD, in September 2010, up to the time of finishing writing

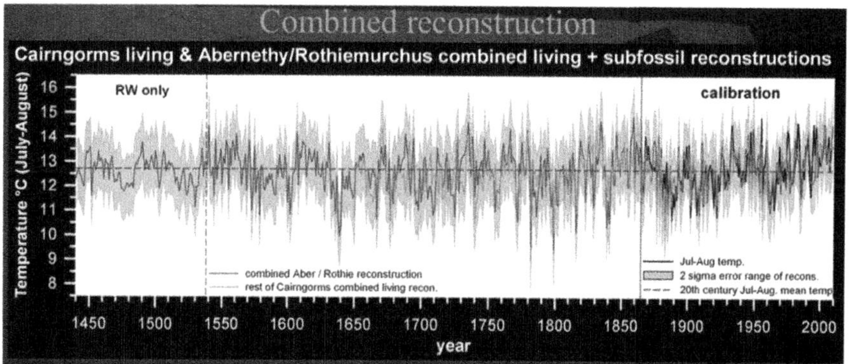

Abernethy-Rothiemurchus vs. Jul-Aug temp.	r^2	DW
1866-2009	51.9%	1.485
Rest of Cairngorms vs. Jul-Aug temp.		
1866-2009	59.1%	1.407
Aber.-Rothie. vs. Rest of Cairngorms		
1866-2009	75.7%	
1650-2009	50.9%	

4.3 Comparing the Cairngorms temperature reconstruction against an 'independent' climate reconstruction from Scotland.

this book in September 2018, I attended nine national (British) and international conferences (I exclude smaller academic meetings like workshops and seminars). These conferences were in the fields of sociology (broadly), history of science, STS, anthropology (broadly), anthropology of climate change and science policy. Attending conferences across such fields in the social sciences forced me to develop an overarching methodology – 'epistemography' – which would be familiar to my diverse group of colleagues, and which would make me feel entitled, if only a little, to speak on behalf of their/our disciplines. In conference talks and Q&A sessions, in corridors, at the bar, and also on the dance floor, I met colleagues with whom I have become emotionally and intellectually close (see González-Santos and Diamond, 2015 for a review of the broader communicative and community-binding functions of conferences). While writing this book, I have imagined the faces of some of these conference-goers and have sought to pre-empt their imagined criticism.

Besides serving the purpose of defining the methodology and the academic audiences of this book, attending conferences allowed me to meet the commissioning editor of this book and to negotiate its content with him. I met the acquisition editor for the subject areas of 'Sociology, Business and

Economics' at Manchester University Press, through a personal contact. I contacted the editor by email prior to the conference (following the advice I read in books about academic publishing) and we agreed to meet in between conference sessions. Back then, I did not have a book proposal, but I prepared a brief one-minute presentation on the book I wanted to write (again, I followed 'self-help' academic books). The editor apparently liked my pitch and invited me to contact him again when I was ready to send him a book proposal. A year later, we met at another conference to discuss the book proposal I had sent him a couple of months earlier. A month after this second meeting, I was offered a book contract. In this way, as Gross and Fleming (2011) argue, academic conferences serve to concretise ideas and move projects forward, as scholars enmesh themselves in webs of interpersonal obligations like agreeing to write a book.

As transcribed below, Hughes's question to Miloš at the conference was about what he called 'the potential self-delusion associated with multiplicity', which Hughes introduced as 'it's probably the kind of thing that we kindly don't mention to our neighbour statisticians because it will be bad for their blood pressure'. The 'multiplicity' or 'multiple comparisons problem' is a typical problem emerging from doing comparability work. As is known in statistics, the multiplicity problem refers to situations when two or more groups are compared using multiple variables (age, education level …). As the number of comparisons increases, it becomes more likely that the objects or groups compared will differ and will render the comparison and the establishment of similarities more difficult. As Hughes made clear, his question about the problems of multiplicity was not only intended for Miloš, which means that comparability work was characteristic of dendroclimatology more generally:

Malcolm Hughes: I have a comment as Hughes et al. 1984 [this is the reference of the paper where Hughes presented the first reconstruction for Scotland. Miloš laughs].

I have a general question that is not aimed exclusively at your presentation. It's something we all get involved in. We try all these different variables; we change this step of the process and that step, and so on. And it's probably the kind of thing that we kindly don't mention to our neighbour statisticians because it will be bad for their blood pressure.

[The entire audience laughs]

Have you got any thoughts about how to deal with the problem of [2–3 seconds' silence] potential self-delusion associated with multiplicity?

[2 seconds' silence]

Miloš: Um, if I understand your question correctly I would say that really … [2 seconds' silence] including many variants and combinations is necessary. It's not possible to derive a single result, an ultimate definitive answer. In that sense, it might be difficult to choose a different approach to develop a reconstruction. [3 seconds' silence]. I am not sure if I've answered your question …

Malcolm Hughes: Can I come back and say that you have come the nearest to answering my question by actually bringing on independent datasets and seeing the same pattern?

Miloš: OK, yes …

Malcolm Hughes: It seems to me that to the extent that they are really independent datasets, that's the way to cut through these issues.

Miloš: I think so too. That's definitely true. You know … Here I have shown that using two different datasets you can achieve more or less the same results or very similar. Yeah, I agree that this is one way … one way to solve this problem. [5 seconds' silence]

Moderator: any other questions?
[A third person in the audience stands up and says]

Attendee: Well, if it was me, if I was talking to a statistician, I'd tell him to go and have a look at the tree physiology library, and to read through all the hundreds and hundreds of studies that relate climate to tree growth. You might get the question [from statisticians] of 'how many observations do you have?' But, you know … it gets a little bit harder when you get six hundred observations to actually go out of the line.

Miloš: Well, yes, what I would say here is that if you want to be sure that the data are showing something real, a real representation of climate, it is necessary to compare it to other regions or areas. In this case, I don't think it is so much of a problem because the instrumental record in the UK or Scotland is quite long and it has been looked at in quite a lot of detail in terms of the quality and so there are different approaches towards how you can validate the data that you've generated and the reconstructions that are coming out of this.

Moderator: Thanks very much. Hopefully, this is the sort of conversation that we can continue over a beer in just one more session.

As Miloš pointed out at the conference another source of evidence that Rob and Miloš employed to evaluate the Cairngorms reconstruction was temperature records and climate reconstructions from other regions in the UK and Europe. More specifically, Miloš compared the Cairngorms reconstruction against the longest temperature record in the world (the Central England Temperature record); Hughes's 1984 Scottish reconstruction; a semi-qualitative temperature reconstruction of Central England based on historical records and created by Hubert Lamb in the 1960s; a Central European temperature reconstruction; an Alpine temperature reconstruction; a Northern Scandinavia temperature reconstruction; a Central Scandinavia temperature reconstruction, and a Pyrenees temperature reconstruction. In the published paper (Rydval et al., 2017: 2962), Miloš represented some of these comparisons in a series of graphs and a table that summarised the intercorrelations between all reconstructions (Figure 4.4).

The last form of comparability work carried out by Miloš and Rob consisted of comparing and triangulating the Cairngorms reconstruction against historical and documentary evidence about extreme weather events in Scotland and major global volcanic eruptions. They interpreted the similarities and differences in relation to broader paleoclimatological debates about twentieth-century climatic change trends. Specifically, paleoclimatologists are trying to ascertain the timing and

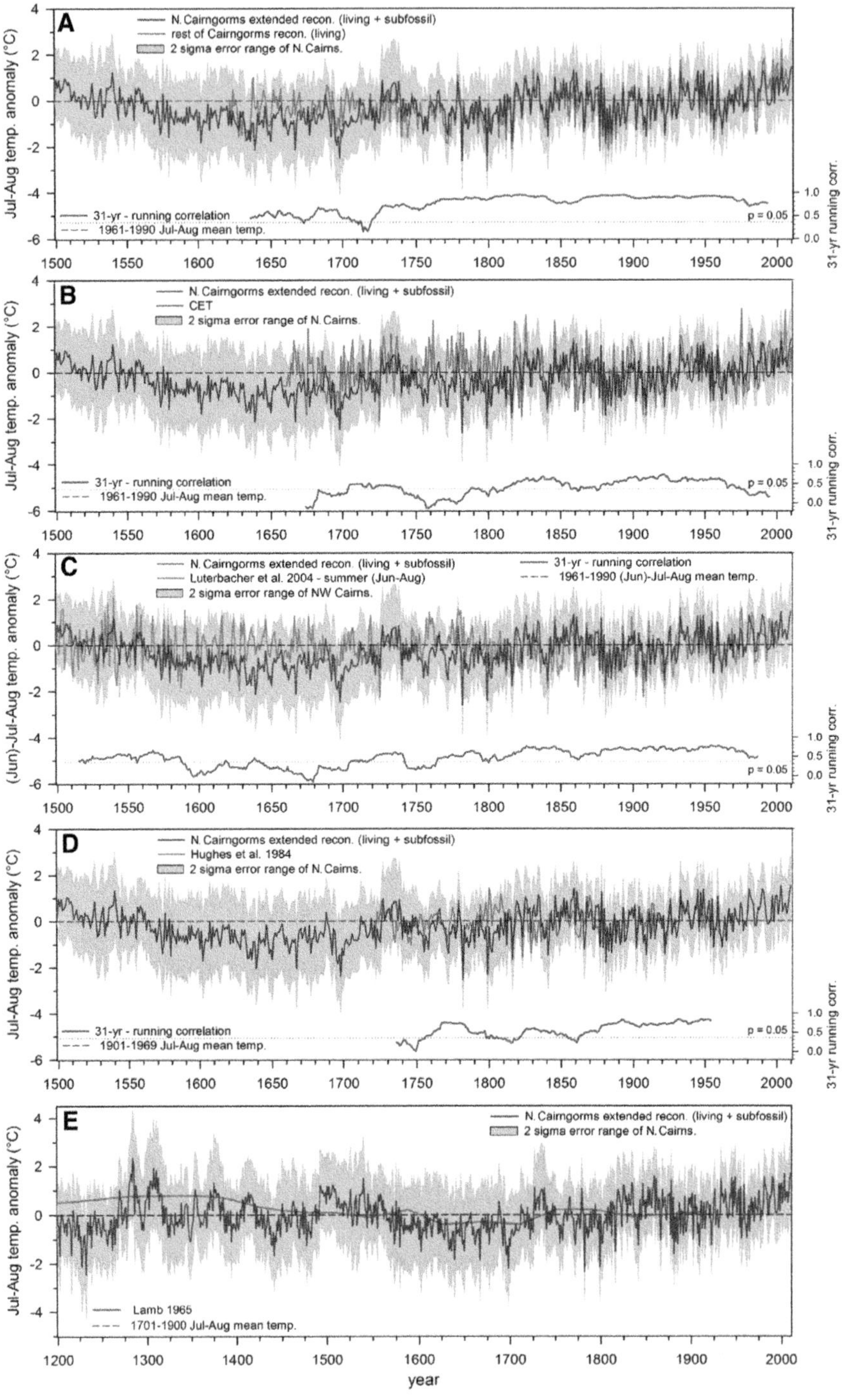
A
N. Cairngorms extended recon. (living + subfossil)
rest of Cairngorms recon. (living)
2 sigma error range of N. Cairns.
31-yr - running correlation
1961-1990 Jul-Aug mean temp.
p = 0.05
Jul-Aug temp. anomaly (°C)
31-yr running corr.
B
N. Cairngorms extended recon. (living + subfossil)
CET
2 sigma error range of N. Cairns.
31-yr - running correlation
1961-1990 Jul-Aug mean temp.
p = 0.05
C
N. Cairngorms extended recon. (living + subfossil)
Luterbacher et al. 2004 - summer (Jun-Aug)
2 sigma error range of NW Cairns.
31-yr - running correlation
1961-1990 (Jun)-Jul-Aug mean temp.
p = 0.05
(Jun)-Jul-Aug temp. anomaly (°C)
D
N. Cairngorms extended recon. (living + subfossil)
Hughes et al. 1984
2 sigma error range of N. Cairns.
31-yr - running correlation
1901-1969 Jul-Aug mean temp.
p = 0.05
E
N. Cairngorms extended recon. (living + subfossil)
2 sigma error range of N. Cairns.
Lamb 1965
1701-1900 Jul-Aug mean temp.
year

4.4a

Table 3 Correlation matrix of NCAIRN reconstruction and other UK temperature records

	NCAIRN	ROC	CET	Luterbacher
1721–1969				
ROC	r=0.867 p<0.001	–	–	–
CET	r=0.400 p<0.001	r=0.418 p<0.001	–	–
Luterbacher	r=0.479 p<0.001	r=0.509 p<0.001	r=0.740 p<0.001	–
Hughes	r=0.489 p<0.001	r=0.546 p<0.001	r=0.363 p<0.001	r=0.467 p<0.001
1659–2002				
ROC	r=0.786 p<0.001	–	–	–
CET	r=0.367 p<0.001	r=0.400 p<0.001	–	–
Luterbacher	r=0.441 p<0.001	r=0.519 p<0.001	r=0.760 p<0.001	–
Gordon castle				
1781–1827	r=0.704 p<0.001	r=0.641 p<0.001	–	–
1879–1974	r=0.722 p<0.001	r=0.690 p<0.001	–	–

4.4b

4.4 Comparing the Cairngorms reconstruction against UK and European reconstructions.

magnitude of two past climatic periods at various locations around the Northern Hemisphere, which are used to contextualise recent global warming: a warm period called 'Medieval Warm Period' (MWP), known to have lasted approximately from 950 to 1250 AD, and a cold period called the 'Little Ice Age' (LIA) spanning from the sixteenth to the nineteenth century. In the article where they first published the temperature reconstruction of Scotland (Rydval et al., 2017: 2957, 2960), Rob and Miloš listed the ten coldest and warmest reconstructed years and the five warmest and coldest reconstructed decades in Scotland (referred to as 'Table 2' in the text below), and noted their coincidences with the MWP and the LIA periods and volcanic eruptions by including in-text citations to published paleoclimate literature that employed tree-rings, documents and instrumental data as sources of evidence:

> Considering the range of uncertainty, the recent warming is not unique with 2001–2010 representing the third warmest decade in the record (Table 2). Other

notably warm reconstructed periods include two shorter periods (1280–1290 and 1300–1320) in the early part of the record, suggesting the possibility of previous warmer conditions during the late Medieval. Other warm decadal periods, similar to the present, are 1490–1510, 1370–1380 and 1730–1740. Despite containing two of the warmest decades (Table 2), the interval around 1300 is associated with considerable uncertainty and reduced replication. Furthermore, representation of this period in the RCS [Regional Curve Standardisation] reconstruction versions may potentially be biased by a concentrated period of recruitment around 1300. However, historical records indicate that the 1280s were marked by climatically favourable conditions with hot, dry summers, though the early 1300s were characterised by deteriorating climate with poor harvests, famine and wet conditions (Dawson 2009; Lamb 1964). It is therefore not clear to what extent this reconstructed early fourteenth century warm period reflects actual climate and this period must therefore be interpreted cautiously at this time.

The most evident extended cold period is centred on the seventeenth century and extends from the late sixteenth until the early eighteenth century (although this is also one of the periods of greatest uncertainty in the reconstruction). This cold period coincides with the so-called Little Ice Age (LIA – Matthews and Briffa 2005) reported in historical and various proxy records from both the Northern and Southern Hemispheres (Büntgen and Hellmann 2014; Neukom et al. 2014) and described as a period of deteriorating climate in Scotland after ~1550 (Lamb 1964). Three of the five coldest reconstructed decades (1631–1640, 1661–1670 and 1691–1700) occurred in the seventeenth century with the 1690s representing the coldest decade in the record (Table 2).The period (~1693–1700) was marked by exceptionally cold and wet summers with widespread famine in Scotland, failed or delayed harvests and southward expansion of sea ice in the northern North Atlantic, coinciding with the effects of volcanic eruptions including the Mt Hekla eruption in 1693 and an unidentified event in 1695 (Dawson 2009; Lamb 1964; Plummer et al. 2012).

In January 2017, Rob and Miloš eventually published the North Cairngorms (NCAIRN) reconstruction in an article called 'Reconstructing 800 years of summer temperatures in Scotland from tree rings' (Rydval et al., 2017) in the journal *Climate Dynamics* (the impact factor of which in 2019 was 3.774 and, according to Journal Citation Reports, 2009, ranked eighth in the list of top 20 journals on climate change). The final temperature reconstruction (Figure 0.1) included two versions. Version 'A' was the 'unfiltered' version of the reconstruction containing all types of climatic trends and frequencies. Version 'B' was a 'low-pass filtered' version, which excluded inter-annual and decadal trends (high-frequency temperature changes) and kept centennial, longer-term temperature changes related to climatic trends. As the article abstract below explains, the final reconstruction had good calibration and verification statistics (the reconstructed temperature values resembled 56.4 per cent of the instrumental data), which were complemented by generally positive results from all the comparability work that Rob and Miloš had carried out:

This study presents a summer temperature reconstruction using Scots pine tree-ring chronologies for Scotland allowing the placement of current regional temperature changes in a longer-term context. 'Living tree' chronologies were extended using 'subfossil' samples extracted from near shore lake sediments resulting in a composite chronology >800 years in length. The North Cairngorms (NCAIRN) reconstruction was developed from a set of composite blue intensity high-pass and ring-width low-pass filtered chronologies with a range of detrending and disturbance correction procedures. Calibration against July–August mean temperature explains 56.4% of the instrumental data variance over 1866–2009 and is well verified. Spatial correlations reveal strong coherence with temperatures over the British Isles, parts of western Europe, southern Scandinavia and northern parts of the Iberian Peninsula. NCAIRN suggests that the recent summer-time warming in Scotland is likely not unique when compared to multi-decadal warm periods observed in the 1300s, 1500s, and 1730s, although trends before the mid-sixteenth century should be interpreted with some caution due to greater uncertainty. Prominent cold periods were identified from the sixteenth century until the early 1800s – agreeing with the so-called Little Ice Age observed in other tree-ring reconstructions from Europe – with the 1690s identified as the coldest decade in the record. The reconstruction shows a significant cooling response 1 year following volcanic eruptions although this result is sensitive to the datasets used to identify such events. In fact, the extreme cold (and warm) years observed in NCAIRN appear more related to internal forcing of the summer North Atlantic Oscillation.

Conclusion

This chapter has detailed the practice of deriving information about past temperature in Scotland from cleaned tree-ring data. The reconstruction of past climates, as I see it, is analogous to the restoration of buildings, mosaics, paintings, manuscripts, old languages and past events. As the social historian Peter Burke (2012: 58) explains, in such cases of restoration, 'much information comes in fragments, and part of the process of the production of knowledge consists in fitting those fragments together as if in a jigsaw puzzle. Such reconstruction or restoration requires knowledge, but it also provides knowledge.' In this way, Burke exemplifies his argument with the case of the German philologist August Schleicher who examined words in present-day Indo-European languages (Germanic, Romanic, Celtic, Slavic, etc.) in order to arrive at their common ancestor, which Schleicher called 'Proto-Indo European.' The fitting together of fragments of the Assyrian tablets and the restoration of mosaics from ancient Roman villas are also reminiscent of the jigsaw metaphor used by Burke. Similarly, as shown in this chapter, the reconstruction of the past climate of Scotland involved working backwards, as in a regress, generating knowledge of a past phenomenon (climate) from the assemblage of present-day evidence (tree-ring data).

The metaphorical description of the work of reconstructing past temperature as a puzzle-solving activity is fitting insofar as, like the assemblage of jigsaws, climate reconstructions are recognised to operate within a system of rules. To solve a jigsaw puzzle is not, for example, to assemble any pieces together and create a random picture. The rules of piecing together a jigsaw puzzle involve interlocking all the pieces, without forcing them, in order to reproduce the cover picture on the puzzle's box. Similarly, reconstructing climate, as shown in this chapter, was not about randomly choosing any climate variable and months of reconstruction. Instead, the rules of reconstructing climate in Scotland involved choosing the most 'limiting' climate variable to tree growth and the target season that 'maximises' the climate signal.

Importantly, the 'rules of play' establishing the adequate resolution of a puzzle or a tree-ring-based climate reconstruction also provide widely admissible problem-solutions or 'moves'. For instance, the puzzle-solver is allowed to group the jigsaw pieces by colour, and Rob and Miloš could take into account useful precedents like Hughes's reconstruction in order to choose the climate variable (temperature) and the target season (July–August). The historian of science Thomas Kuhn (1970 [1962]: 42) – who also used the analogy with puzzles to describe ordinary scientific practice – referred to these accepted solutions as 'paradigms' or 'strong network of commitments', which one acquires by virtue of being a member of a community. He explained:

> The existence of this strong network of commitments – conceptual, theoretical, instrumental, and methodological – is a principal source of the metaphor that relates normal science to puzzle-solving. Because it provides rules that tell the practitioner of a mature specialty what both the world and his science are like, he can concentrate with assurance upon the esoteric problems that these rules and existing knowledge define for him. What then personally challenges him is how to bring the residual puzzle to a solution. In these and other respects, a discussion of puzzles and of rules illuminates the nature of normal scientific practice.

Another revealing aspect of the parallelism between resolving puzzles and reconstructing past climates is that the resolution of such tasks is modelled on available solutions but is not resolved entirely by them. To resolve a puzzle for the first time, the neophyte might draw upon his or her knowledge of similar games like chess, but this familiarity will not suffice because puzzle-solutions are different from chess-solutions. The more experienced puzzle-solver will draw on conventional 'moves' such as grouping the pieces by colour, starting the puzzle from the edges and so on, but resolving a new puzzle of 3,000 pieces will always require some ingenuity. Likewise, the strategy of trained variation and natural selection that Rob and Miloš employed to resolve the problem of reconstructing the climate in Scotland partly drew upon an accepted precedent (Hughes's reconstruction) but also required an ingenious solution for choosing the 'best' reconstruction (the combo approach). The existing approach ('Esperism') did not automatically resolve, in Rob's eyes, the challenge of having a more 'objective' selective method. Rob's role, as an experienced dendroclimatologist or puzzle-solver, was precisely to point out the inadequacies of existing solutions and suggest a new way forward.

The analogy with puzzles is misleading, however, if we think that the temperature reconstruction of Scotland was independently validated by calibration and verification statistics, like when a puzzle is completed by comparing it with the cover picture on the box. Precisely, Hughes's question to Miloš during the Melbourne conference about the problem of multiplicity, as I interpret it, was a warning against naively accepting a certain quantitative threshold as sufficient evidence of the accuracy of climate reconstructions. Hughes's insistence on using 'independent' datasets, and the multiple forms of comparability work carried out by Miloš and Rob, point to the importance they gave to forms of evidence and reasoning other than statistical in interpreting the validity of reconstructions. By triangulating different paleoclimate data about extreme climatic years and periods (from tree-rings, documents and instrumental data), Rob and Miloš effectively mobilised different bodies of paleoclimatological knowledge for the purpose of validating the reconstruction from Scotland (see Sturdy, 2007 for an example of triangulating practices in medicine). In a way, the broader community of paleoclimatology to which Miloš and Rob belonged became implicated in the verification of the climate reconstruction of Scotland.

The work of reconstruction described in this chapter illustrates, more broadly, how new scientific knowledge (including epistemographic knowledge; see Box 7) develops from existing knowledge by way of analogy and induction. In the words of the sociologist Barry Barnes (1982: 52), 'By seeing the unknown in terms of a known problem-solution, inductive inference is possible: variables in the unknown situation are calculated by assuming that it behaves analogously to a known one. Thus science proceeds by analogy and induction, with the former licensing the latter. Where analogy is perceived, expectation is projected.' This view of knowledge formation is called 'finitism', the core perception of which, as Barnes (1982: 30) explains, is that 'proper usage [of a concept] is developed step by step, in processes involving successions of on-the-spot judgements. Every instance of use, or of proper use, of a concept must in the last analysis be accounted for separately, by reference to specific, local, contingent determinants.' According to finitism, the meaning of concepts is open-ended because concepts have been employed a finite number of times (hence its name), and when individuals encounter a new item they have to negotiate whether this item is sufficiently similar to the previous items classified using that concept. For instance, consider the term 'murder', and how existing cases defined as 'murder' do not suffice for all possible applications of the term in the present, which is why there are still debates about whether 'murder' includes the killing of enemy soldiers, human foetuses, animals or terminally ill people who have expressed a wish to be helped to die (Hatherly et al., 2008: 7).

Analogously, climate reconstructions are finite epistemic objects, because, as dendroclimatologists know very well, they are likely to be revised as new tree-ring data are generated. In dendroclimatology, the cognitive and material framework that enables analogies between known present climate (in the calibration/verification periods) and unknown past climate (beyond meteorological records) is the statistical method of linear regression and the 'principle of uniformitarianism', which, as one textbook author admits, is a necessary 'productive starting point' in creating tree-ring based climate reconstructions. The role of imaginings, aspirations and expectations in science should, therefore, not be underestimated. Scholars from so-called 'expectation studies' (Borup et al., 2006: 293) have argued that 'expectations play a central role in science and technology not least because they mediate across boundaries between different scales, levels, times and communities'.

Precisely, the way dendroclimatologists employ uniformitarianism as a hopeful 'productive starting point' serves the purpose of binding claims about present, past and future climate change. As Skrydstrup (2017) argues, paleoclimatology is also concerned with foretelling the future, as it seeks to reconstruct past climates in order to anticipate future climate scenarios (a form of reasoning that he aptly calls 'analogue anticipation'). In this way, historical reconstructions of climate become analogues or puzzle-solutions against which model simulations of future climate are evaluated. As the author of one of the main paleoclimatology textbook explains (Bradley, 1999: 1), 'Paleoclimate data provides the basis for testing hypotheses about the causes of climate change. Only when the causes of past climatic fluctuations are understood will it be possible to fully anticipate or forecast climatic variations in the future.' As a twist to the uniformitarianism slogan attributed to James Hutton ('the present is the key to the past'), I would argue that the work of climate reconstructions is, therefore, sustained on the hope that 'the past is the key to the present and the future'.

The conclusion that knowledge about historical climates proceeds analogically and creatively is important for public debates about the formation and nature of scientific consensus. The public's notion of scientific consensus (including scientists' own views) can be characterised with the metaphor of the 'wall of bricks', which suggests that scientific knowledge (the wall or the whole building) is built up in small incremental additions of new truths to the stock of old truths (new bricks on old bricks) by many individuals and groups over time (Altman, 2012). The metaphor of the wall of bricks was behind many of the responses of climate scientists after the hacking of the climate science emails. For example, some of the authors of the 2008 IPCC report rebuffed the allegations of misconduct in the IPCC peer-review process by depicting consensus-building as a matter of piling up results: 'The body of evidence is the result of the careful and painstaking work of hundreds of scientists worldwide. The internal consistency from multiple lines of evidence strongly supports the work of the scientific community, including those individuals singled out in these email exchanges' (Revkin, 2009).

At the core of the wall of bricks metaphor there is the assumption that scientific consensus results from the accumulation of papers, datasets and multiple forms of evidence. Surveys attempting to measure the level of scientific consensus concerning climate change from the analysis of article abstracts typically espouse this accumulative view of knowledge (see Bray, 2010 for a review of such studies). According to the 'wall of bricks metaphor', once the consensus has been painstakingly built, only very ingenious and brave scientists can tear it down. This is why in a letter signed by 225 climate scientists after the hacking of climate science emails, the theory of global warming was equated to 'well established' theories and 'facts' such as the theory of evolution, the Big Bang theory and the theory of the origin of Earth: 'Even as these [theories] are overwhelmingly accepted by the scientific community, fame still awaits anyone who could show these theories to be wrong. Climate change now falls into this category' (The Guardian, 2010a).

In contrast to the popular view of scientific consensus as a solid and progressive approximation to an ultimate truth, this chapter suggests an alternative view of consensus as an elusive achievement, always in process of formation in talk, labour and sociality spaces participated in by relatively small communities of experts who tentatively agree that existing problem-solutions (theories, methods, datasets) resolve, partially or totally, new problems. We should accept that scientific facts have a transient rather than an absolute character, otherwise we will reach the self-defeating sceptical conclusion that scientists cannot know at all, as many past scientific theories have now been dismissed and present theories will likely be abandoned or reformulated in the future. How can we explain the emergence and potential resolution of a research problem like the 'divergence problem' or the abandonment of the theory of global cooling – which predicted a twentieth-century cooling of the Earth's surface – if not by accepting the dynamic nature of scientific knowledge? If Rob and Miloš were uncertain about the magnitude of recent climate change in Scotland given all the paleoclimate evidence, why would we ask them to provide more definitive conclusions? If one misunderstands scientific consensus as a matter of progressively achieving unanimity in a group, how can one not perceive the normal to and fro between scientists like Rob and Jan Esper with regards to the choice of the 'best' temperature reconstruction as a challenge to knowledge? If we lump climate scientists into two polarised groups (those who agree with the 'consensus position' of the IPCC and those who disagree), where do Rob and Miloš's nuanced findings fit in? Given the specialisation and fragmentation in climate research (Weart, 2018), why are we surprised about the existence of conflicting perspectives in the study of climate?

Some might legitimately worry that the alternative image I propose of scientific consensus might hinder political action on climate change. The relevant literature, however, shows that the degree of scientific consensus about an issue often has little influence on policy action (Pearce et al., 2017; Sarewitz, 2004). If we understand that the relationship between science and policy is not linear, it will be easier to accept that scientific disagreements and uncertainties do not inherently harm the authority of science and policy (in fact, as Shackley and Wynne, 1996 show, it can lead to quite the contrary).

5

Controversy

Rob and Miloš produced the temperature reconstruction of Scotland against the background of a scientific controversy. This discussion originated when Michael E. Mann (mentioned in the preface), Jose D. Fuentes and Scott D. Rutherford (hereafter 'Mann et al. (2012)') published an article in February 2012 in the journal *Nature Geoscience* formulating the 'missing-ring hypothesis'. Essentially, this hypothesis explained the supposed problems that Mann et al. (2012) had identified in a Northern Hemisphere temperature reconstruction that Rob had co-authored with Professor Rosanne D'Arrigo and Professor Gordon Jacoby (this reconstruction is abbreviated using the initials of the authors' surnames and the year of publication: 'DWJ06'; D'Arrigo et al., 2006).

Throughout my fieldwork, Rob and Miloš implicitly referred to this controversy by naming, often sarcastically, the lead author of Mann et al. (2012) (see Box 9). For instance, while sampling a lake during my first fieldwork expedition in August 2012, Rob told me, 'You see? This is the difference between people like Mann who sit at a desk and use archived tree-ring data to formulate stupid hypotheses, and those like us who create the data.' Likewise, Rob occasionally mocked the fact that Mann posed in his university website with wood samples, 'I doubt Mann has ever done any sampling or crossdating.' As a lead author of DWJ06, Rob became the corresponding author in responding to Mann et al. (2012). However, because Rob had employed tree-ring data from other colleagues to generate DWJ06 (see Chapter 2 on the dendrochronological practice of creating tree-ring chronologies) the controversy eventually involved the broader community of dendroclimatology.

Box 9

Dealing with 'missing out' syndrome

As Rob kept mentioning Mann's name at different times during my fieldwork I slowly realised that I was 'missing out' on the opportunity of generating data on something important that was happening while I was busy investigating something else (see Reynolds, 2017 for further reflection on the feelings of 'missing out' widely shared among ethnographers). When I first heard Rob talking about Mann in August 2012, I had just started my research and I did not know who Mann was. I wrote down Rob's ironic comment in my field notes nonetheless, because I realised it was important to him, and I placed a question mark next to Rob's words. At the time, I was struggling to understand what was going on in the fieldwork expeditions at the Scottish Highlands, and I did not follow up Rob's comment about Mann. As my research progressed, I heard Rob and Miloš occasionally referring to Mann, but, again, I was studying other aspects of their work and I just made a note of it. As I could not follow the controversy as it was happening, I asked Rob to 'keep me updated'. For instance, I learnt about Montford's incendiary blog post because Rob told me about it. Similarly, I was able to keep up with the numerous articles that dendroclimatologists published in response to Mann et al. (2012) thanks to Rob's help and my own follow-up research of journal articles, online forums and blogs.

My approach to the reconstruction of the 'missing ring' controversy is exemplary of the way I dealt with the difficulty of researching simultaneous phenomena. 'Fieldwork', 'dendrochronology', 'standardisation' and 'reconstruction' were concurrent epistemic practices, in the sense that the decisions that Rob and Miloš made at certain stages were conditioned by their knowledge of problems at later stages. For instance, Rob's decision to focus the 2014 fieldwork expedition on producing more subfossil samples was driven by the problem of 'low replication' he had encountered earlier while crossdating chronologies. Likewise, Rob and Miloš's decision to use the tree-ring data from the East of Scotland (the Cairngorms) for the reconstruction was motivated by their previous discovery of disturbance in the broader dataset. My realisation of the interlinkages between practices emerged over time, as I researched and wrote about such practices linearly and fragmentally, and I managed to get a sense of the overall circle of epistemic work.

I organised my fieldwork akin to what Jeffrey and Troman (2004: 538) describe as 'a compressed time mode', which in my case involved generating successive and time-limited observations of practices and feeding back ideas which, only in retrospect, became intelligible. For instance, when I spent a few months studying the work of dendrochronology (because I saw that Miloš was focused on doing such work), I was able to understand, and wrote down, comments that Miloš made about fieldwork, which was the only practice I had previously studied (for instance, I noted that Miloš often complained of twisted cores that came out of bad fieldwork). Similarly, when I attended the conference in Melbourne to generate data about the last stage of the reconstruction (Rob and Miloš had attended this conference specifically to present the reconstruction), I was also attentive to any supplementary observations I could generate about the previous practices (for instance, that Miloš won the prize for his standardisation work).

Michael E. Mann is currently a Distinguished Professor of Atmospheric Sciences at Pennsylvania State University, and is well known in paleoclimatology and broader circles of science as the creator of the so-called 'hockey-stick graph', a reconstruction of the average Northern Hemisphere temperature over the past 1,000 years that shows a slow cooling trend to 1900 (the hockey-stick's 'shaft') followed by a sharp increase of temperatures in the late twentieth century (the 'blade' portion of the stick). Mann co-authored the paper about the hockey-stick graph with Raymond Bradley, the author of the main paleoclimatology textbook, and Malcolm Hughes, the creator of the first Scottish reconstruction (Mann et al., 1998).

The hockey-stick graph became an icon of twentieth-century global warming when the IPCC included it in the 2001 Summary for Policy-makers, and Mann became a highly popular and prolific scientist. Besides publishing more than 200 peer-reviewed papers, he has written four books on climate science, the latest of which is a children's book. He has also been awarded numerous prizes; most significantly, the 2007 Nobel Peace Prize, which was jointly awarded to the former US vice-president Al Gore Jr and to thousands of experts that had participated in the IPCC reports (including Mann himself). Mann is also an influential public figure, both in the US and around the world: he has been asked to testify twice in the US Congress; has written numerous op-eds and commentaries; has been invited to give various 'public outreach' talks'; is the co-creator of a popular science blog ('Real Climate'), and, at the time of writing this text (September 2018), has more than 70,000 Twitter followers. As a result of his popularity, Mann's work has also been the focus of intense public criticism, as he recounts in his autobiographical book *The Hockey Stick and the Climate Wars: Dispatches from the Front Lines* (Mann, 2012).

In their controversial article 'Underestimation of volcanic cooling in tree-ring based reconstructions of hemispheric temperatures', Mann et al. (2012) inferred the existence of missing rings in the tree-ring chronologies that Rob had used to create DWJ06 by noting a 'glaring inconsistency' between DWJ06 and a model-based climate prediction of past volcanic cooling. Mann et al. (2012) had used two climate models (a coupled ocean–atmosphere general circulation model and an energy-balance climate model) to simulate the temperatures after four volcanic eruptions known to have occurred in 1258, 1452, 1809 and 1815. Mann et al. (2012: 203) reported that 'Both models predict a drop of 2° Celsius following the 1258/1259 eruption, whereas the reconstruction [DWJ06] shows a decrease of only 0.6° Celsius. A similar pattern holds for the two other largest eruptions.'

Crucially, Mann et al. (2012) attributed the mismatch to a deficiency in DWJ06 rather than in the climate models. They explained, 'Given their success in reproducing volcanic cooling events of the historical era, we might expect the models' predictions for previous centuries to be similarly reliable' (Mann et al., 2012: 202). In other words, as climate models were deemed to simulate the Krakatau 1883 volcanic cooling recorded by thermometers better than DWJ06, Mann et al. (2012) assumed that DWJ06 had also 'underestimated' the previous past volcanic coolings in 1258 and 1452. Crucially, this conclusion relied entirely on two assumptions (that instrumental data are the 'golden standard' and that the principle of uniformitarianism upholds) which, as seen in Chapter 4, also underlined the creation of the tree-ring-based temperature reconstructions of Scotland.

Mann et al. (2012) employed a variant of the same tree-growth model (VS) that Rob and Miloš had previously used to confirm the existence of disturbance in the Scottish dataset (Chapter 3). Mann et al. (2012) used the VS model to generate hypothetical evidence of missing rings in DWJ06; they generated modelled chronologies that represented the effect of volcanic cooling on trees growing at Northern latitudes. Mann et al. (2012) offered a physiological explanation of their hypothesis: the volcanic particles released into the air blocked some sunlight, causing cooling and a diffuse light at the surface, which in turn caused trees at the Northern Hemisphere to stop growing and to produce missing rings for the years after the volcanic eruptions. Accordingly, Mann et al. (2012) adjusted the original parameters to the growth conditions after a volcanic eruption in the same way that Miloš had adjusted the VS-Lite parameters to the wet conditions in Scotland. For instance, Mann et al. (2012) established that any tree-ring that the VS produced for periods of growth below 26 days could be interpreted as a missing ring. Yet, Mann et al. (2012: 202) claimed, rather controversially, 'Our findings are insensitive to the precise details of the growth model.'

The epistemological conundrum that dendroclimatologists, and Rob in particular, faced was to deal with a relatively close colleague who challenged fundamental claims and practices of their community. Specifically, Mann et al. (2012) raised doubts about the practice of purposively sampling certain areas and about the possibility that samples from the Northern Hemisphere could be systematically biased as they could underestimate volcanic cooling. Likewise, Mann et al. (2012) suggested that dendrochronologists could not detect missing tree-rings with the method of crossdating, and hence could not produce properly dated tree-ring chronologies. Finally, the missing-ring hypothesis also raised questions about the reliability of extrapolating information about past climate from tree-ring data given that certain tree-ring chronologies would not reflect volcanic cooling as estimated by climate models. In this chapter I explain Rob's active role in publicly defending and securing support and acceptance – first from his close community of colleagues and later from members of the blogosphere – of tree-ring-based climate reconstructions (including the Scottish one) as accurate accounts of historical changes in climate.

The 'community' responds

Rob became particularly implicated in galvanising his colleagues to write what he referred to as the 'community response'. From Rob's perspective, the missing-rings hypothesis did not only challenge his work but also all Northern Hemisphere reconstructions and the entire field of dendrochronology. Shortly after Mann wrote a post in his blog 'Real Climate' announcing the publication of Mann et al. (2012),

Rob wrote a comment (Wilson, 2012b) on behalf of twelve people whose forenames he included (possibly indicating that Mann knew them personally):

> Dear Mike,
> Your paper has certainly generated a lot of discussion over the last few days between some dendroclimatologists.
>
> You would appreciate that we are somewhat sceptical of your hypothesis and analyses and are drafting an appropriate measured response to your work.
>
> With kind regards,
> Rob, Rosanne, Ed, Kevin, Keith, Tom, Jan, Dave, Ulf, Brian, Håkan and Paul

After a few email exchanges, Rob eventually gathered a total of twenty-three signatories, most of whom he knew personally and had previously worked with on tree-ring-based climate reconstructions from the Northern Hemisphere. More specifically, Rob had co-authored papers with twelve out of the total twenty-three signatories (Kevin J. Anchukaitis, Keith Briffa, Ulf Büntgen, Edward Cook, Rosanne D'Arrigo, Jan Esper, David Frank, Björn E. Gunnarson, Malcolm Hughes, Paul Krusič, Brian Luckman and Thomas M. Melvin), and four of them were involved, directly or indirectly, in the Scottish Pine Project (Kevin Anchukaitis, Edward Cook, Björn Gunnarson and Malcolm Hughes). The twenty-three co-authors were all based at universities in North America and Europe (the United States, the United Kingdom, Canada, Germany, Austria and Russia). The list included four women (see Copenheaver et al., 2010 for an analysis of the gender co-authoring practices in dendroclimatology) and three of the five most cited dendroclimatologists (Edward Cook, Kevin Anchukaitis and Eugene Vaganov) according to Google Scholar (2019). After a peer-review process that Rob qualified as 'unduly long', their response was published online on 25 November 2012 – nine months after Mann et al. (2012) – in *Nature Geoscience*, with the title 'Tree rings and volcanic cooling' (hereafter 'Anchukaitis et al. (2012)').

Anchukaitis et al. (2012) disputed the missing-ring hypothesis on three grounds. First, Anchukaitis et al. (2012) accused Mann et al. (2012) of selecting 'arbitrary' and 'unrealistic' parameters in the VS model. Among other problems, they emphasised, 'Mann and colleagues arbitrarily and without justification require 26 days with temperatures above their unrealistic threshold for ring formation … These assumptions all bias Mann and colleagues' tree-growth model results towards erroneously producing missing tree rings' (Anchukaitis et al., 2012: 836). As Miloš pointed out to me, this criticism was invested with extraordinary authority as the creators of the VS model (Vaganov–Shashkin) were among the co-authors and 'they know what they are talking about'. Secondly, Anchukaitis et al. (2012) criticised Mann et al. (2012) for not considering the possibility that climate models were to blame for the mismatch with instrumental data. They cited a few uncertainties of climate models, noting that 'the timing and magnitude of cooling in climate model simulations is uncertain', which they regarded as evidence that 'an alternative hypothesis of an overestimation of volcanically induced cooling in the simulations cannot be ruled out' (Anchukaitis et al., 2012: 837). Finally, Anchukaitis et al. (2012: 836) criticised Mann et al. (2012) for 'a lack of any empirical evidence of misdating errors in tree-ring chronologies'. Instead, Anchukaitis et al. (2012) provided 'empirical evidence' that tree-ring data were accurately dated by creating a graph that showed a synchronicity between an 'independently-produced' tree-ring density chronology for the Northern Hemisphere – which had not been used in DWJ06 – and the timing of explosive volcanic eruptions as recorded by thermometers (see Chapter 4 for a similar credibility strategy of comparing against 'independent' tree-ring chronologies).

On the same day that Anchukaitis et al. (2012) was published online, Rob sent a message to the members of a public online forum, the 'International Dendrochronology Discussion list' (ITRDBFOR), announcing its publication. Rob presented the implications of the missing-tree-rings hypothesis as a matter of concern not only to dendroclimatologists, but also to the entire community of dendrochronologists as 'Their main hypothesis was that there was a temporary cessation of tree growth (i.e. missing rings for all trees) at some sites near the temperature limit for growth. This implies Dendrochronology's inability to detect missing rings' (Wilson, 2012c). In response, a member of the online forum ironically asked, '"A temporary cessation of tree growth" resulting in no rings for all trees? Now this is a hypothesis that I am willing to bet good money has no empirical support since studies of trees began 200 years or so ago. Speculation this bold could give dendrochronologists a bad name' (Lanner, 2012). In response to this latter message, Malcolm Hughes added that 'No dendrochronologists were involved in the offending Mann et al 2012 paper. What Rob described was the response of a number of us to some of the multiple flaws in the original paper' (Hughes, 2012).

As the online dendrochronology forum was and still is publicly accessible, bloggers immediately reported on the publication of Anchukaitis et al. (2012) and speculated about the nature of such controversy. The creator of the blog 'Watts Up With That?' wrote the post 'Dendros stick it to the Mann' and noted that a couple of dendroclimatologists who had previously worked with Mann were now among the twenty-three signatories: 'What is most interesting is that Hughes and Briffa are co-authors of the response to Mann' (Watts, 2012). Similarly, Andrew Montford, the blogger with whom Rob had occasionally collaborated (Chapter 3) also wrote a post with the title 'Lonely Old Mann' and commented, 'The list of authors of the new paper is very long. Almost looks like they are ganging up on him.;-)' (Montford, 2012).

In response to Montford's post, Rob wrote a blog comment where he sought to clarify his disagreement with Mann. He lamented that Mann had 'wasted a lot of time for many people' as Mann had not discussed his hypothesis with some of his dendrochronology colleagues nor looked at 'some real tree-ring data to learn what "crossdating" is'. In Rob's own words: 'Mann's major flaw was to see something in his model which did not agree with "nature" and assumed that there must be something wrong with nature. Alas, if he had taken the trouble either (1) to speak to some of his dendrochronological colleagues or (2) look at some real tree-ring data to learn what "crossdating" is, he would have quickly realised that his hypothesis was wrong and would not have wasted a lot of time for many people' (Wilson, 2012a).

The online controversy

In October 2013, almost a year after Anchukaitis et al. (2012) was published, Rob became embroiled in an online discussion that started after Montford reported in his blog 'Bishop Hill' on Rob's lecture on millennial temperature reconstructions to which Montford had been invited. Montford started his post published on the 21st of October by introducing Rob: 'As readers here know, Rob is no kind of a sceptic (a point he repeated over lunch), but on the northern hemisphere paleo studies his position is not a million miles away from mine. In places our positions are identical, as you will see.' Montford followed by describing Rob's lecture as a dire criticism of Mann's work: 'The real fireworks came when Mann's latest papers, which hypothesise that tree ring proxies have large numbers of missing rings after major volcanic eruptions, were described as "a crock of xxxx [shit]" (Montford, 2013). In a few hours, Montford's blog post triggered dozens of comments. Mann also reported on it in Twitter (Figure 5.1), first seemingly accusing Rob of being a 'climate denier' (Mann deleted this tweet) and then accusing him instead of 'not be[ing] a climate denier but play[ing] a contrarian role in debate' (Mann, 2013a) and asking him to 'disavow' Montford's blog post (Mann, 2013b).

Mann's tweets about Rob triggered some reactions in social media, particularly from 'Bishop Hill' readers, who largely praised Rob. One blog reader expressed 'Huge Respect to Rob' (flaxdoctor, 21 October 2013, 9.40am); another reader stated that 'My own view is that you [Rob] have done climate science a major service, as have Curry and few others, by being prepared to take an independent line and consider critically the meanings and implications of the research that has been done and probably is still in train' (mikejackson, 21 October, 5.07pm); another one thanked Rob 'for being willing to vet this post for our host, taking the time to comment here and especially for making the relevant literature more accessible' (frank, 21 October 2013, 0.07pm); and one last criticised Mann's characterisation of Rob as a 'climate change denier' as 'Mann's tweet just reveals openly what has long been his working assumption. To Mann, a "skeptic" is anyone who doesn't accept his work uncritically, and a "denier" is anyone who actually disagrees with him' (McKitrick, 2013). Among those who tweeted in support of Rob there was a British computer modeller and a popular climate science communicator who, in reference to Mann's deleted tweet, asked him: '[Y]ou are seriously calling Rob a denier for criticising your work, M? That's pretty strong to call a prof climate colleague' (Edwards, 2013) and a dendrochronologist who tweeted 'Rob Wilson (U St. Andrew's) is a fine dendrochronologist and paleoclimatologist, a thoughtful scientist, and 100% not a "climate denier"' (St George, 2013).

Rob's participation in Montford's blog also raised some concerns by participants in social media. The author of the blog '...and Then There's Physics' (2013) criticised Rob for 'engaging in contrarian sites' and 'doing so in a way that ultimately just seems to provide more ammunition for them to then use to undermine the scientific evidence', and one blog reader asked Rob 'to disassociate himself with the BH [Bishop Hill] blog, unless it clarifies its position' (toby52, 26 October 2013, 1:44pm). In Twitter, the director of the NASA Goddard Institute for Space Studies and

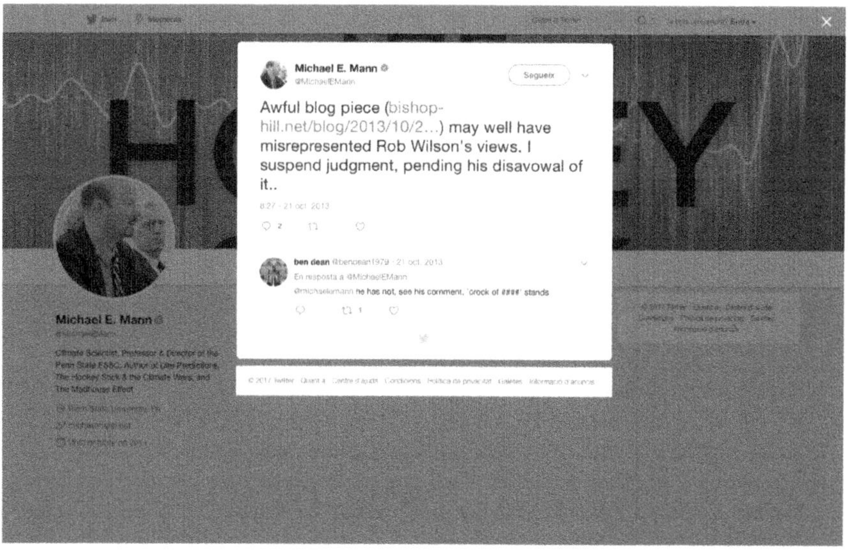

5.1a

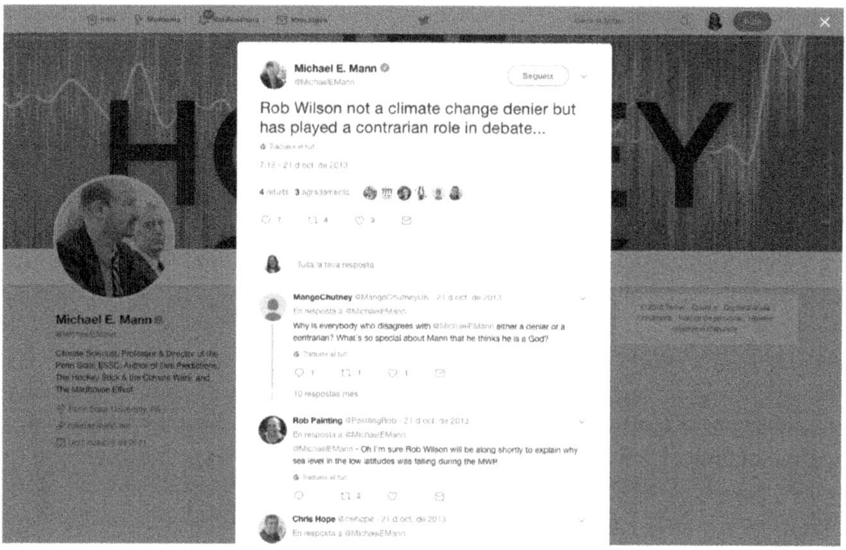

5.1b

5.1 Mann's online response to Rob entrenched the controversy even more.

co-editor with Mann of the blog 'RealClimate', implicitly criticised Rob when he wrote, 'Science is not linear. Interesting ideas can be proposed & challenged (w/o anyone's work being a "crock"). Leads to deeper understanding' (Schmidt, 2013).

On the same day that Montford's blog was posted, Rob wrote a comment where he clarified that 'my 2hour lecture was … not focussed entirely on Michael Mann's work' and insisted that '[although] the "crock of xxxx" statement was ... rather flippant, I stand by it'. The comment also included a list of thirteen open-source articles thematically classified by 'Northern Hemisphere-related papers' and 'Missing tree-rings and major volcanic events' (Wilson, 2013):

> Although I vetted Andrew's post, I want to clarify that my 2 hour lecture was, I hope, a critical look at all of the northern hemispheric reconstructions of past temperature to date. It was not focussed entirely on Michael Mann's work. I described each of the major studies and tried to highlight both their strengths and weaknesses – they all have some useful information but it is important to understand the limitations of the studies as well. Of course Mann's work was mentioned as several of his papers have been so prominent over the last 15 years but I actually spent substantially more time taking apart the D'Arrigo et al. (2006) study on which I did much of the analysis.
>
> This was a session where I wanted the students to critically look at the different studies and specifically address what we can learn from them and how the science can move on over the next decade. Such large scale reconstructions are critically important for understanding the controls on large climate variability, but as yet, due to great uncertainties and large differences in reconstructed amplitude, they are not yet very useful at constraining modelled estimates of future temperature change.
>
> Bar some personal comments, much of what I said is published (see papers below) and is in the public domain.
>
> Lastly, the 'crock of xxxx' statement was focussed entirely on recent work by Michael Mann w.r.t. hypothesised missing rings in tree-ring records (a whole bunch of papers listed below). Although a rather flippant statement, I stand by it and Mann is well aware of my criticisms (privately and through the peer reviewed literature) of his recent work.

The twitter and blog spats continued for a few days (the last comment on Montford's blog was on 26 October). A cartoonist, who published regularly on Montford's blog and is the creator of the 'Climate Skeptics Calendar' created a cartoon that drew upon Rob's 'crock of xxxx' statement to criticise Mann's previous work on the 'hockey-stick graph' (Josh 241 (2013)). Rob followed the online exchanges closely and told me he felt somewhat 'anxious' about the possibility that Mann could ostracise him from the community of climate scientists, as 'Mann is a very influential scientist' (at that point I was also worried that my research could also make this conflict worse, see Box 10). Rob was also disappointed about the fact that he had received one email from Mann saying that he considered his professional and personal relationship with Rob finished. 'The sad thing is that I've never got to meet Mike and talk about all this with him', Rob lamented.

Box 10

Is this epistemography objective and useful?

I was and I am still concerned that my analysis of the controversy over the missing-ring hypothesis could be discredited for being biased and supportive of Rob and dendroclimatologists (this concern is well founded, as other epistemographers have become 'captives' in controversies; see Collins, 1996; Scott et al., 1990). My analysis of the controversy is inevitably partial because my sources of data have been filtered by Rob (Box 9). Likewise, my analysis is certainly one-sided, as I portray this controversy from the point of view of Rob and his colleagues. Portraying Mann's or Montford's positions accurately would have required conducting an in-depth study of paleomodelling or the blogosphere similar to the one I did about dendroclimatology (an option that I discarded not only because I lacked time, but also because, given my friendship with Rob, I doubt I could have ever gained Mann's trust).

As I see it, the unavoidable partiality of my research can hardly be criticised. Analysing the controversy from a 'view from nowhere' – which does not report on any particular position – or a 'view from everywhere' – which reports on all the positions – was impossible, and therefore I believe it is unfair to assess the objectivity of my analysis on such unviable standards. As an alternative, I offer a new criterion for evaluating the objectivity of my epistemographic analysis. In this way, rather than discussing whether my epistemography is objective and useful (on someone else's terms and conditions), I seek to demonstrate that it is indeed objective and useful given my own definition (here I draw upon Daniel Neyland's considerations of ethnographic utility, 2008: 166–174).

As I see it, producing an objective analysis of any given controversy would imply showing the achievement of inter-subjectivity and agreement within the parties of the dispute in a way that coherently accounts for the reality of the subjects' experience and for comparable controversies studied by disciplinary colleagues (in defining objectivity in such terms I partly draw upon Alan Fine and Hallett, 2014). In other words, epistemographers have to consider the expectations of objectivity and utility upheld by their two main audiences: academic peers and subjects ('funding bodies' and 'society' are increasingly important audiences). On the one hand, my epistemography colleagues value the creation of theory. New theory should resonate not only with my field observations, but also with existing theories that colleagues have developed from other epistemic cultures and controversies (Box 7). On the other hand, my subjects value their credibility and reputation. Rob and Miloš still do not know if this book will help them to become more credible to sceptical audiences, but they hope it will. This is why, when I once asked Rob to publicly reflect upon the value of my work for him, he responded in the form of a question, 'Does being a sociologist's "case study" provide a new avenue of science communication and outreach?' (Ramírez-i-Ollé, 2019a: 308).

With this book I seek to reconcile the demands of objectivity and utility from my subjects and academic peers. In the conclusion section of this chapter, for instance, I have outlined the broader socio-political significance of the missing-ring controversy – by drawing upon and thus validating the research and theories developed by my peers – and have brought this analysis to bear on public discussions about the nature of scientific scepticism, which are of relevance to climate scientists. More generally, in the conclusion sections of this book, I have used epistemographic knowledge to draw the 'public issues' from the 'personal troubles' experienced by Rob and Miloš; much in the same way that Charles Wright Mills characterised the social analyst's skill as 'the capacity to shift from one perspective to another – from the political to the psychological; from examination of a single family to comparative assessment of the national budgets of the world; from considerations of an oil industry to studies of contemporary poetry. It is the capacity to range from the most impersonal and remote transformations to the most intimate features of the human self – and to see the relations between the two' ([1959] 2000: 7).

Like Wright Mills, I am convinced that the 'sociological imagination' that epistemographers display in their work can become useful to their subjects and readers. Wright Mills explains, 'It is by means of the sociological imagination that men now hope to grasp what is going on in the world, and to understand what is happening in themselves as minute points of the intersections between biography and history within society' ([1959] 2000: 7). In other words, Wright Mills believed that an awareness of one's position in history and society can ease one's personal troubles and transform the course of public issues. As stated in the introduction, I also wish to challenge damaging myths about climate science ('public issues') and to ease climate scientists' anxiety about their credibility ('personal troubles').

The (temporary) closure of the controversy

While Anchukaitis et al. (2012) was being reviewed in *Nature Geoscience*, dendrochronologists (many of whom were also co-authors of the 'community response') were writing up other studies for specialised journals of dendrochronology, volcanology and geophysics in which they tried to finally disprove the claims made by Mann et al. (2012). The choice of specialised journals is significant insofar as, in the opinion of the dendroclimatologists and dendrochronologists I talked to, the controversy had been publicised too widely. One dendroclimatologist told me that Mann has been 'disrespectful towards the [dendrochronology] community in airing his damaging hypothesis in a widely read journal like *Nature Geoscience* where other scientists might get a wrong impression from our work'. In his opinion, it would have been more appropriate if Mann had published his hypothesis in a dendrochronology journal so that 'we would have dealt with it first'. One dendrochronologist reckoned that the publication of Mann et al. (2012) had had an overall positive effect on dendrochronology because 'If there's something we have to thank Mann for, it's to force us to demonstrate the good work we've done for decades.'

In their follow-up articles, dendrochronologists re-examined tree-ring data in the light of the missing-rings hypothesis. In May and June 2013, a group of German and Swiss dendrochronologists reported the results of their re-examination of the longest density chronology from Northern and Central Europe and concluded that 'the MAN12 hypothesis [Mann et al., 2012] on post-volcanic missing rings can be rejected based on simple comparisons of tree-ring, instrumental and documentary data over the past 300–500 years from Central and Northern Europe' (Esper et al., 2013a). In another paper, these same researchers disregarded Mann et al. (2012)'s suggestion that volcanic cooling could have an effect on long-term climatic trends (Esper et al., 2013b). In July 2013, three North American dendrochronologists published the analysis of thousands of ring-width chronologies from the Northern Hemisphere archived in the International Tree-Ring Data Bank and calculated a 'percentage of the frequency of missing rings'. They concluded that the missing-rings hypothesis 'is not consistent with the pattern of absent-ring formation outlined by more than 17 million tree rings. Locally absent rings are extremely rare in tree-ring records from high latitudes and high elevations' (St George et al., 2013: 3730).

With two other colleagues, Rob co-authored a paper (D'Arrigo et al., 2013) that examined recent dendroclimatological reconstructions from the Northern Hemisphere where they brought to bear the temperature reconstruction from Scotland, among other datasets. In particular, they employed the tree-ring chronology from the Cairngorms to refute Mann's hypothesis that 90 per cent of the tree-ring chronologies from the Northern Hemisphere would have a missing ring in 1815, the year of a volcanic eruption. They created a graph that showed the agreement between ring-width and density datasets from Scotland against the long monthly temperature series for Scotland over a long period, including the 1815 volcanic

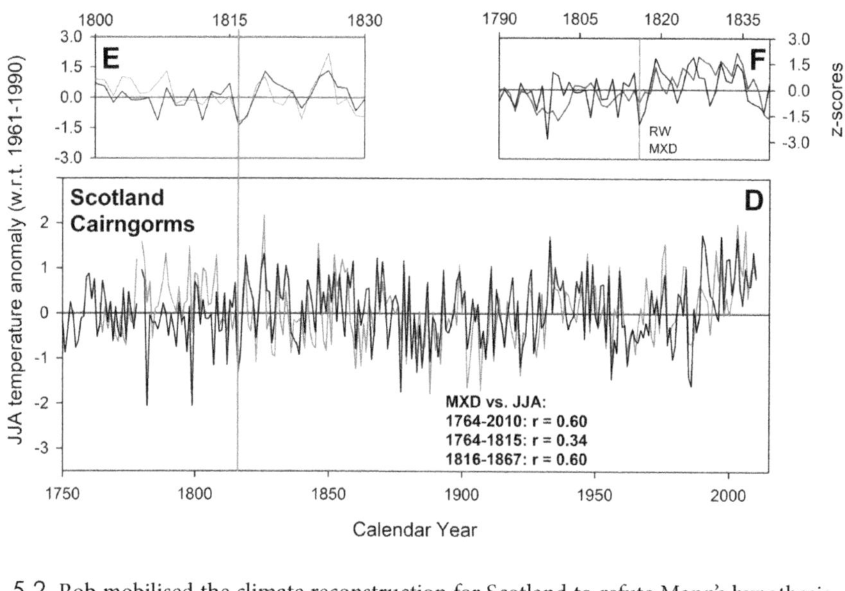

5.2 Rob mobilised the climate reconstruction for Scotland to refute Mann's hypothesis.

year. This synchrony was interpreted by Rob and colleagues as evidence of the inexistence of a missing tree-ring because 'if the 1816 tree ring was missing from all Scottish trees then the correlation would break down. For the 52 year period from 1816 to 1867, MXD [density] correlates with JJA [July–August] mean temperatures at 0.60. Prior to 1816 (1764–1815), the correlation is 0.34' (D'Arrigo et al., 2013: 7).

Co-authors in D'Arrigo et al. (2013) also drew on Rob's knowledge of the 'disturbance' and environmental history of Scotland to explain that the density chronologies from the Cairngorms showed a lower correlation response than ring-width data (0.34 compared with 0.60) prior to the 1815 volcanic eruption. 'Although the correlation between MXD [density] and growing season climate is weaker for this earlier period, this likely reflects both the markedly weaker replication in the MXD [density] records in the 18th century as well as management related disturbance in these woodlands through the 18th and early nineteenth-century (Wilson et al., 2012)' (D'Arrigo et al., 2013: 7). They concluded by restating the value of the method of crossdating over the modelling techniques used in Mann et al. (2012): 'Given the past century of the proven methodology of crossdating in dendrochronology, the MFR12a theory [Mann et al. (2012)] can only be validated by using evidence from real tree ring data rather than model simulations (which have been shown to not accurately reflect tree biology or the actual distribution of the DWJ06 network' (D'Arrigo et al., 2013: 9).

In May 2014, a group of European dendrochronologists ('Büntgen et al., 2014') claimed in the generalist journal *Nature Climate Change* to have found a potentially conclusive refutation of the missing-ring hypothesis. More specifically, Büntgen et al. (2014) used the radiocarbon dates of a ring-width chronology from the Alps (also included in DWJ06) as an 'independent' confirmation of the dating of all the remaining tree-ring chronologies from the Northern Hemisphere (see Chapter 2 for the radiocarbon-dating method). The authors reported that the Alpine chronology showed an increase in concentration of carbon isotopes from the year 774 to the year 775, which coincided in time with a peak in carbon isotopes in two ring-width chronologies from Japan and Germany respectively. They concluded that the fact that the carbon date AD 775 coincided in three chronologies from different continents offered 'an independent, geochemical age determination for dendrochronologically dated tree-ring chronologies' (Büntgen et al., 2014: 404), and thus a confirmation that the chronologies used in DWJ06 were accurately dated. They concluded the article by suggesting the possibility that a reassessment of existing carbon-dated tree-ring chronologies around the year 775 would definitely foreclose the controversy (the results of this study were eventually published in a 2018 article co-authored by forty-seven people which had the telling title 'Tree rings reveal globally coherent signature of cosmogenic radiocarbon events in 774 and 993 CE'; Büntgen et al., 2018).

In a response published in July 2014, Michael Mann and Scott Rutherford (hereafter 'Rutherford and Mann, 2014') accepted radiocarbon-dating 'as an independent time-marker necessary to directly test our hypothesis' (Rutherford and Mann, 2014: 649). Rutherford and Mann (2014) used the 774–775 radiocarbon date analysed by Büntgen et al. (2014) as a standard against which to test a new simulated chronology they had generated with their VS model, adjusted to simulate the effects of Alpine cold conditions on tree growth. They reported that their simulated chronology and the Alpine chronology were similar, which was interpreted as a confirmation of both the hypothesis of missing rings and the claim made by Büntgen et al. (2014) that the Alpine chronology was well dated. For Rutherford and Mann, however, the ultimate way of refuting their hypothesis was to simulate all the ring-width chronologies used in DWJ06 that reached back to AD 774, and compare them against carbon-dated chronologies. In this way, they agreed with Büntgen et al. (2014) that further (carbon-dating) analyses of tree-ring data would confirm if the missing-ring hypothesis was correct.

The controversy continued in peer-review journals for a couple of years as dendroclimatologists published new studies where they sought to refute further the missing-ring hypothesis (Anchukaitis et al., 2017; Wilson et al., 2016; Schneider et al., 2015; Stoffel et al., 2015). The two protagonists of the controversy, Rob Wilson and Michael Mann, seemed to have lost interest in engaging further in the debate. After mounting a 'community response' and publishing multiple 'revisionist' studies, including the Büntgen et al. (2018) paper, Rob felt that there was sufficient evidence to foreclose the controversy. While Mann regarded the controversy open until new radiocarbon-dating evidence was made available, Rob perceived Mann's latest work (Mann et al., 2013) as 'awful' and advised me to 'ignore it'. To this date, Mann has not published any reply to the latest dendroclimatological papers (including Büntgen et al., 2018); a silence that Rob interprets as indicative that 'Mann knows he is wrong – just will not admit it.'

Conclusion

This last empirical chapter has described a debate within contemporary climate research over the accuracy of tree-ring-based reconstructions of the Earth's changing climate. The dynamics of the controversy over the missing-ring hypothesis exemplify some of the characteristics of scientific disputes occurring in the twenty-first century. While the official responses were published in standard forums of scientific discussion such as peer-reviewed journals, the development of the controversy on blogs, Twitter and email correspondence illustrates the existence of new places for debate in science and the effects that the Internet, the blogosphere and the instantaneous transmission of news and opinion might have on entrenching disagreements (this conclusion is in line with recent social research that has examined the role of social media in political and social conflicts; Zeitzoff, 2017). At the same time, I suggest, the controversy on the missing-ring hypothesis is also typical of past controversies within climate science.

Historiographical and anthropological research on climate science has shown the deep socio-cultural roots of ongoing tensions between two broad scientific approaches to studying climate (Heymann et al., 2017; Edwards, 2010; Heymann, 2010). On one hand, there is an older, empirical-descriptive, fieldwork-based approach based on geology, geography, archaeology and cognate qualitative-orientated disciplines, which seeks to understand climate change in its historical, geographical and cultural context. On the other hand, there is a more recent, deductive, quantitative, computerised and model-based approach developing from the mid-twentieth-century convergence of the disciplines of atmospheric sciences, statistics and mathematics, which regards climate change as a physical and law-regulated phenomenon. The epistemic conflict between the 'empirical' and 'modelling' approaches is reflected in the professional careers of distinguished climate scientists such as Gordon Manley (Endfield et al., 2015); Hubert Lamb (Martin-Nielsen, 2015) and Frederick Seitz (Lahsen, 2013a: 739), who were dissatisfied with the way numerical modelling grew to dominate climate research and who lost professional relationships, funding and jobs as a result of the diminished credibility of their empirical approach. As Demeritt (2001: 315) points out, these historical conflicts were 'not simply a case of petty in-fighting; it was also about what kind of science would be practiced and how knowledge would be evaluated'.

As I see it, the controversy over a relatively minor technical issue such as the missing-ring hypothesis represents a major socio-political conflict between two cultures within climate research (the 'empirical' and 'modelling') over what constitutes the best way of knowing about historical climate change, and which applications of that knowledge are politically more useful and worth investing in by society. On the one hand, dendroclimatologists like Rob, with their emphasis on observations generated from proxy sources, represent the empirical approach. On the other hand, paleomodellers like Mann give greater prominence to the evidence generated by computer and deductive simulations in line with the modelling approach. The three types of criticism formulated in the 'community response' encapsulate the fundamental disagreements between dendroclimatologists and paleomodellers. For the twenty-three signatories, Mann et al. (2012) had proposed a hypothesis with no biological and theoretical soundness as their simulated tree-ring data were, in the words of Mann et al. (2012: 204), 'insensitive to the precise details of the growth model'. As Mann et al. (2012) did not have experience working with 'real tree-ring data', dendroclimatologists believed that Mann et al. had chosen 'arbitrary' and 'unrealistic' parameters, and wrongly disregarded crossdating as a method for generating appropriate 'empirical' evidence.

It is important to recognise that participants in the controversy shared some assumptions and language that enabled their disagreements and conversation in the first place. For instance, they all regarded meteorological data as the most credible source of evidence against which to evaluate their simulations and reconstructions of past temperatures, and they also both effectively accepted uniformitarianism as a valid assumption in the study of past climate. Similarly, the fact that Mann et al. (2012) and Büntgen et al. (2014) trusted the carbon-dating method as an indisputable source of verification of tree-ring-dating might be a way of potentially closing the controversy for all sides. Therefore, I must insist that cultural differences between dendroclimatologists and paleomodellers do *not* necessarily lead to conflict. As Ribeiro Duarte (2017) has shown, these differences are routinely managed through everyday interaction. My own ethnographic observations confirm this latter point. In September 2016, just before the latest phase of the project funding of the Scottish Pine Project expired, Rob organised a meeting with close collaborators to discuss the results of the project. Among the small group of fourteen colleagues, Rob invited two paleomodellers whom he casually referred to as 'outsiders' and 'end-users'. In his presentation, Rob insisted that these two modellers should not 'black box' the Scottish data, and extensively discussed the uncertainties of the reconstruction from Scotland (focusing on the disturbance problem; see Chapter 3). In turn, these two modellers asked questions about how the tree-ring data had been produced and standardised, and joked that 'We don't even know what a tree is because we always sit in front of the computer.'

An important factor that explains, I think, why there was a clash between 'epistemic cultures' (Knorr Cetina, 1999) in the case of the controversy over the missing-ring hypothesis is that participants had different visions of what experts were competent to evaluate the hypothesis. In other words, dendroclimatologists and dendrochronologists more broadly saw Mann et al. (2012) as not only inaccurate but also offensive as, in the words of one of them, Mann had been 'disrespectful towards the community' because he had not allowed the appropriate experts (dendrochronologists) to examine his hypothesis. As Rob expressed on Montford's blog, Mann had not spoken to 'some of his dendrochronological colleagues' and had wasted their time. On the online forum, Hughes also related his offence to the fact that 'No dendrochronologists were involved in the offending Mann et al 2012 paper'. By publishing their 'revisionist' articles in the specialised journals of dendrochronology, volcanology and geophysics, dendroclimatologists exemplified the kind of audiences and experts to whom they thought Mann should have addressed his hypothesis. As a result of their own appraisal, dendrochronologists reaffirmed not only their own sceptical practices, but, crucially, their shared body of knowledge; as one put it, 'If there's something we have to thank Mann for, it's to force us to demonstrate the good work we've done for decades.

By offering his hypothesis to the sceptical examination of an audience beyond the dendroclimatology and dendrochronology community, Mann strained, if not totally jeopardised, the trusted position in which he was held by its members and, as bloggers were quick to point out ('the Dendros stick it to the Mann' and 'Lonely Old Mann'), dendroclimatologists readjusted Mann's membership of their community of trusted peers. For many dendroclimatologists, Mann evolved from being seen as a trusted colleague with whom to produce relatively accurate paleoclimatological knowledge like the 'hockey stick' to being considered an unreliable collaborator, and even an outsider, whose hypothesis had to be responded to collectively.

The group of twenty-three signatories constitute, I suggest, an example of the importance of collegial and friendship relationships in the arts and sciences. The sociologist Michael P. Farrell (2003) calls 'collaborative circles' the intertwinement of creative and friendship dynamics he observed, among others, in the French Impressionists, Sigmund Freud and his friends, C. S. Lewis, J. R. R. Tolkien and the Inklings. In turn, the sociologist of science Harry Collins (1988; 1981) identified the key role of 'core-sets', or small groups of experts who function on the basis of interpersonal trust and familiarity, in foreclosing any given scientific controversy. As Collins (1981: 8) insists members of core-sets do not need to be friends and often disagree with each other to the extent that 'some members may be enemies, and this can lead to complete lack of communication between them, except via third parties, in formal settings'. For instance, as we know from Chapter 4, Rob and Jan Esper (two of the signatories of the 'community response') disagreed over the best way of choosing a reconstruction. Collins (1981: 12) also insists that core-sets are ephemeral and 'transient hot-spots in science' as scientists are often engaged in other types of work at the same time as participating in a core-set, as is the case with Rob and the Scottish Pine Project. Finally, Collins (1988: 740–741) also suggests that the boundaries of core-sets change as a result of processes of 'over-restrictiveness' – when access into a scientific debate is controlled and restricted – and 'over-extension', when relatively untrained and inexperienced individuals are included in expert discussions.

The controversy over the missing-ring hypothesis illustrates the specific dynamics and consequences associated with the tightening and extension of the boundaries of an elite group of experts. Because they thought he displayed incompetent forms of reasoning and evaluation, the twenty-three signatories restricted Mann's access to dendroclimatological matters. Importantly, Mann's exclusion from the narrow circle of Rob's close friends and collaborators might be temporary, and under different circumstances Mann might collaborate again with them. More evidently, Mann is trusted by broader circles of science, as he has received many scientific awards since the start of the controversy in 2012. At the same time, the extension of the core-set to members of the blogosphere effectively closed the controversy for Rob and Mann who broke relations and stopped conversing. By engaging with outsiders whom Mann considered to be unreliable, Rob became seen as, if not a climate denier, at least 'play[ing] a contrarian role in debate'. Rob's anxiety about being ostracised from the broader circles of science where Mann was seen to be influential also shows that a single person can simultaneously be an 'insider' and an 'outsider', depending on one's membership of different 'core-sets' or communities of trust and knowledge (Merton, 1972).

Rob's relationship with Montford and his blog readers and the participation of the blogosphere in the controversy demonstrate that dendroclimatology, and climate science more generally, are more permeable to the scepticism of outsiders than some of these outsiders themselves might imagine. In the aftermath of the hacking of climate science emails, some bloggers and journalists criticised climate scientists for being an exclusive and close-minded group. For instance, Mosher and Fuller (2010: 8) lamented that 'We are tough on the scientists we call The Team, and we think deservedly so. But we want to stress from the outset that we do not for one minute believe there is any evidence of a long term conspiracy to defraud the public about global warming, by the Team or anyone else. What we find evidence of on a much smaller scale is a small group of scientists too close to each other, protecting themselves and their careers, and unintentionally having a dramatic, if unintended effect on a global debate'. Likewise, a *Guardian* (2010b) editorial complained that the stolen emails showed how some climate scientists manoeuvred to prevent critics whom they did not trust from publishing in journals: 'The settled core of our knowledge on climate – the fact of increasing atmospheric carbon, the rising temperature trend, and the heat-trapping mechanism linking the two – has acquired the terrific authority it now possesses precisely because it has been forced to withstand so many challenges in the past. The moment climatology is sheltered from dispute, its force begins to wane. So the sort of closing of intellectual ranks witnessed at UEA [University of East Anglia] was serious and, in the end, self-defeating.'

Admittedly, Rob is exceptional in the extent to which he was willing to establish relationships with outsiders to the community of climate science (including myself), as some of his colleagues qualified Rob's interactions as 'crazy' and 'risky', and one of them even recognised Rob's exceptionality in terms of 'Rob is kinda our representative in the sceptical world'. Rob is well aware of the perception of his peers, 'as they think I'm nuts', but he rationalised his behaviour as: 'My motivation comes down to frustration that individuals don't get what I think is obvious so I try to interact and persuade'. In this way, Rob's self-reported motivation to interact with sceptical outsiders was to re-educate them on the issues that he considered important.

The fact that Rob's trusting attitude might be unusual among climate scientists is even more important for appreciating that his attempts to persuade people whom his colleagues would generally consider 'outsiders' effectively resulted in Rob adopting the outsiders' scepticism and integrating it as an internal evaluative standard in the community, like a 'boomerang-effect' process (Hovland et al., 1953) that I call 'inside-out scepticism'. Unlike most members of his community, Rob trusted outsiders like McIntyre and Montford because he thought that, for the most part, they provided useful scepticism and could contribute to making dendroclimatology more robust. Besides the controversy analysed in this chapter, other examples of inside-out scepticism throughout the book refer to situations when Rob invoked McIntyre, Montford or the 'sceptics' as an imagined public in lectures, conferences and conversations with students and colleagues. In Chapter 3, I recounted how Rob advised Miloš to think about how he could 'rationalise' their standardisation choices to 'sceptics' like Montford. Similarly, Rob asked his undergraduate students to respond to the criticism of Montford's blog readers as part of one of their assignments. In Chapter 4, we saw that Rob's 'combo approach' was motivated, as he explained at the Melbourne conference, by the fact that he wore 'McIntyre's hat' and sought to pre-empt McIntyre's criticism with regards to an 'objective' method for choosing the 'best' reconstruction. One last example of inside-out scepticism was Hughes's 'devil's advocate' type of question to Miloš during the Melbourne conference, where he mentioned the 'problem of multiplicity' as a potential criticism from outsiders (statisticians). All these examples indicate how in the interest of making dendroclimatological knowledge more credible to critical outsiders, Rob and some of his colleagues integrated the outsiders' scepticism and used it to examine each other's claims and practices.

The paradox that therefore emerges from this book is that people who are commonly considered to be 'outsiders' to climate science might be much more involved, if indirectly, in the making of dendroclimatological knowledge than supposed 'insiders' like Mann, whose expertise and potential contribution to dendroclimatology, as it stands, is perceived critically by many dendroclimatologists. This conclusion is important for public and academic debates that conceive debates about climate change as fitting into homogeneous and stereotyped groups of 'climate sceptics', 'deniers' or 'contrarians'. Such labels are not simply descriptive but are also performative categories that individuals use to structure their relations with others (as argued by labellist theorists such as Spector and Kitsuse, [1977] 2001; Becker, 1963). For instance, Mann used the labels of 'climate denier' or 'contrarian' to delegitimise and exclude Rob from the community of people whom he thought could be regarded as reliable colleagues (and this is why Rob felt that he was being ostracised). Similarly, bloggers and their readers diverged over whether Rob was 'no kind of sceptic' or if, instead, Rob had 'to disassociate himself with the BH blog', and such labelling served to create boundaries among themselves and within the blogosphere. Unfortunately, a few social scientists still use the labels of 'climate sceptics', 'deniers' or 'contrarians' unreflexively, as an explanation or description of existing behaviour rather than as a starting point for analysis, and hence have contributed to polarising public debates about global warming (see Howarth and Sharman, 2015 and Lahsen, 2013b: 550 for a similar critique).

I have argued elsewhere (Ramírez-i-Ollé, 2018) and insist again here, backed up with new evidence from this book, that the identity of a 'climate sceptic' or any 'uncivil sceptic' – that is, someone whose scepticism is perceived as inappropriate – is relative to the eye of the beholder and to the sceptic's contingent trust relations with the members of a community of peers who trust each other as decorous experts.

Bibliography

Agar, Jon (2012) *Science in the Twentieth Century and Beyond*. London: Polity Press.

Alan Fine, Gary and Hallett, Tim (2014) Stranger and stranger: creating theory through ethnographic distance and authority. *Journal of Organizational Ethnography*, 3 (2): 188–203.

Almklov, Petter G. and Hepsø, Vidar (2011) Between and beyond data: how analogue field experience informs the interpretation of remote data sources in petroleum reservoir Geology. *Social Studies of Science*, 41 (4): 539–561.

Altman, Douglas (2012) Building a metaphor: another brick in the wall? *BMJ: British Medical Journal*, 345 (7888): 17–18.

Anchukaitis, Kevin J., Breitenmoser, Petra, Briffa, Keith, Buchwal, Agata, Büntgen, Ulf, Cook, Edward, et al. (2012) Tree rings and volcanic cooling. *Nature Geoscience*, 5 (12): 836–837.

'... and Then There's Physics' (2013) Engaging with sceptics. 26 October, https://andthentheresphysics.wordpress.com/2013/10/26/engaging-with-skeptics/, accessed 15 August 2018.

Anchukaitis, Kevin, Wilson, Rob. J., Briffa, Keith R., Büntgen, Ulf, Cook, Edward R., D'Arrigo, Rosanne, et al. (2017) Last millennium Northern Hemisphere summer temperatures from tree rings: part II: spatially resolved reconstructions. *Quaternary Science Reviews*, 163: 1–22.

Arat, Alp (2017) We need to talk about mindfulness. 'LSE Religion and the Public Sphere', http://blogs.lse.ac.uk/religionglobalsociety/2017/04/we-need-to-talk-about-mindfulness-the-changing-face-of-religion-and-the-secular-in-the-public-sphere/, accessed 8 September 2018.

Ashmore, Malcolm (1989) *The Reflexive Thesis: Wrighting Sociology of Scientific Knowledge*. Chicago: University of Chicago Press.

Atkinson, Paul (1992) *Understanding Ethnographic Texts*. Newbury Park; London: Sage Publications.

Baillie, Michael (2010) Tree-ring patterns are intellectual property, not climate data. *The Guardian*, www.theguardian.com/environment/2010/may/11/climate-science-tree-ring-data, accessed 8 September 2018.

Baillie, Michael (1995) *A Slice through Time: Dendrochronology and Precise Dating*. London: Routledge.

Balmer, Brian (2012) *Secrecy and Science: A Historical Sociology of Biological and Chemical Warfare*. Farnham: Ashgate.

Barnes, Barry (2016) On 'The Social Construction of Reality': reflections on a missed opportunity. *Human Studies*, 39 (1): 113–125.

Barnes, Barry (2013) On social constructivist accounts of the natural sciences. In M. Carrier, J. Roggenhofer, G. Küppers, et al. (eds) *Knowledge and the World: Challenges beyond the Science Wars*. Berlin: Springer, pp. 105–136.

Barnes Barry (2005) The credibility of scientific expertise in a culture of suspicion. *Interdisciplinary Science Reviews*, 30 (1): 11–18.

Barnes, Barry (2001) Practice as collective action. In Theodore R. Schatzki, Karin Knorr Cetina and Eike von Savigny (eds) *The Practice Turn in Contemporary Theory*. London: Routledge, pp. 25–36.

Barnes, Barry (1985) *About Science*. Oxford: Basil Blackwell.

Barnes, Barry (1983) Social life as bootstrapped induction. *Sociology*, 17 (4): 524–545.

Barnes, Barry (1982) *T. S. Kuhn and Social Science*. London: Macmillan.

Barnes, Barry and Bloor, David (1982) Relativism, rationalism and the sociology of knowledge. In Martin Hollis and Steven Lukes (eds) , *Rationality and Relativism*. Cambridge, MA: MIT Press, pp. 21–47.

Barnes, Barry, Bloor, David and Henry, John (1996) *Scientific Knowledge: A Sociological Analysis*. Chicago; London: University of Chicago Press.

Bea, Sara (2017) *No heroics, please: mopping deceased donation practices in a Catalan hospital*. PhD thesis, The University of Edinburgh. www.era.lib.ed.ac.uk/handle/1842/23376.

Beaulieu, Anne (2010) Research note: from co-location to co-presence: shifts in the use of ethnography for the study of knowledge. *Social Studies of Science*, 40 (3): 450–470.

Beck, Silke, Forsyth, Tim, Kohler, Pia M., Lahsen, Myanna and Mahony, Martin (2016) The making of global environmental science and politics. In Clark A. Miller, Rayvon Fouché, Laurel Smith-Doerr and Ulrike Felt (eds) *The Handbook of Science and Technology Studies*. Cambridge, MA: MIT Press, pp. 1059–1086.

Beck, Ulrich (1992) *Risk Society: Towards a New Modernity*. London: Sage.

Becker, Howard (1998) *Tricks of the Trade: How to Think About Your Research While You're Doing It*. Chicago; London: University of Chicago Press.

Becker, Howard (1963) *Outsiders: Studies in the Sociology of Deviance*. New York: Free Press of Glencoe.

Berger, Peter and Berger, Brigitte (1972) *Sociology: A Biographical Approach*. New York: Basic Books.

Berger, Peter and Luckmann, Thomas (1966) *The Social Construction of Reality: A Treatise in the Sociology of Knowledge*. London: Penguin Books.

Berger, Peter and Zijderveld, Anton (2009) *In Praise of Doubt: How to Have Convictions Without Becoming a Fanatic*. New York: HarperOne.

Biagioli, Mario (2006) *Galileo's Instruments of Credit: Telescopes, Images, Secrecy*. Chicago: University of Chicago Press.

Bloor, David (2013) Performative theory of institutions. In B. Kaldis (ed.) *Encyclopedia of Philosophy and the Social Sciences*. Thousand Oaks: Sage Publications, Inc., pp. 706–709.

Bloor, David (2008) Relativism at 30,000 ft. In Massimo Mazzotti (ed.) *Knowledge as Social Order: Rethinking the Sociology of Barry Barnes*. Aldershot: Agate, pp. 13–33.

Bloor, David ([1976] 1991). *Knowledge and Social Imagery*. Chicago: Chicago University Press.

Blumer, Herbert (1954) What is wrong with social theory. *American Sociological Review*, 18 (1): 3–10.

Borup, Mads, Brown, Nik, Konrad, Kornelia and Van Lente, Harro (2006) The sociology of expectations in science and technology. *Technology Analysis and Strategic Management*, 18 (3–4): 285–298.

Bouldin, Jim (2012) Severe analytical problems in dendroclimatology, part two. 'Ecologically Orientated', 24 November, https://ecologicallyoriented.wordpress.com/2012/11/24/severe-analytical-problems-in-dendroclimatology-part-two/, accessed 8 September 2018.

Bourdieu, Pierre (1986) The forms of capital. In J. Richardson (ed.) *Handbook of Theory and Research for the Sociology of Education*. Westport, CT: Greenwood, pp. 241–258.

Bourdieu, Pierre and Wacquant, Loïc (1992) *An Invitation to Reflexive Sociology*. Chicago: University of Chicago Press / Polity.

Bowker, Geoffrey C. and Leigh Star, Susan (1999) *Sorting Things Out: Classification and Its Consequences*. Cambridge, MA: MIT Press.

Bradley, Raymond (1999) *Paleoclimatology: Reconstructing Climates of the Quaternary*. Amsterdam; London: Academic Press.

Braun, Erik (2017) Mindful but not religious: science and naturalized enchantment in the work of Jon Kabat-Zinn. In David McMahan and Erik Braun (eds) *Buddhist Studies and the Scientific Study of Meditation*. New York: Oxford University Press.

Bray, Dennis (2010) The scientific consensus of climate change revisited. *Environmental Science Policy*, 13 (5): 340–350.

Briffa, Keith (1984) *Tree-Climate Relationships and Dendroclimatological Reconstruction in the British Isles*. PhD thesis, University of East Anglia.

Brisset, Nicolas (2017) The future of performativity. *Œconomia – History/ Methodology/ Philosophy*, 7 (3): 439–452.

Büntgen, Ulf, Wacker, Lukas, Galván, J. Diego, Arnold, Stephanie, Arseneault, Dominique, Baillie, Michael, et al. (2018) Tree rings reveal globally coherent signature of cosmogenic radiocarbon events in 774 and 993 CE. *Nature Communications*, 9, 3605.

Büntgen, Ulf, Wacker, Lukas, Nicolussi, Kurt, Sigl, Michael; Güttler, Dominik, et al. (2014) Extraterrestrial confirmation of tree-ring dating. *Nature Climate Change*, 4 (6): 404–405.

Burke, Peter (2012) *A Social History of Knowledge II: From the Encyclopédie to Wikipedia*. Cambridge: Polity.

Buroway, Michael (2005) Combat in the dissertation zone. *American Sociologist*, 36 (2): 43–56.

Butterworth, Jon (2010) Come on 'philosophers of science', you must do better than this. *The Guardian*, 17 March, www.theguardian.com/science/blog/2010/mar/17/philosophy-science-climate-change, accessed 8 September 2018.

Callon, Michel (1998) Introduction: the embeddedness of economic markets in economics. In Michel Callon (ed.) *The Laws of the Markets*. Oxford; Maldon: Blackwell Publishers / The Sociological Review, pp. 1–57.

Chang, Hasok (2004) *Inventing Temperature: Measurement and Scientific Progress*. Oxford: Oxford University Press.

Cicerone, Ralph (2010) Ensuring integrity in science. *Science*, 327 (5966). doi:10.2307/40510196.

Collins, Harry (2014) *Are We All Scientific Experts Now?* Cambridge; Malden: Polity.

Collins, Harry (2004a) How do you know if you've alternated? *Social Studies of Science*, 34 (1): 103–106.

Collins, Harry (2004b) *The Gravity's Shadow: The Search for Gravitational Waves*. Chicago: Chicago University Press.

Collins, Harry (2001) Tacit knowledge, trust, and the Q of Sapphire. *Social Studies of Science*, 31 (1): 71–85.

Collins, Harry (1996) In praise of futile gestures: how scientific is the sociology of scientific knowledge? *Social Studies of Science*, 26 (2): 229–244 .

Collins, Harry (1988) Public experiments and displays of virtuosity: the 'core-set' revisited. *Social Studies of Science*, 18 (4): 725–748.

Collins, Harry (1985) *Changing Order: Replication and Induction in Scientific Practice*. London; Beverly Hills: Sage Publications.

Collins, Harry (1981) The place of the 'core-set' in modern science: social contingency with methodological propriety in science in innovation and continuity in science. *History of Science*, 19 (1): 6–19.

Collins, Harry M. and Pinch, Trevor ([2002] 2014) *The Golem at Large: What You Should Know about Technology* (6th ed.). Cambridge: Cambridge University Press.

Collins, Harry M. and Pinch, Trevor ([1993] 2012) *The Golem: What You Should Know about Science* (2nd ed.). Cambridge; New York: Cambridge University Press.

Collins, Harry M. and Pinch, Trevor (2005) *Dr. Golem: How to Think about Medicine.* Chicago: University of Chicago Press.

Cook, Edward (1987) The decomposition of tree-ring series for environmental studies. *Tree-Ring Bulletin*, 47: 37–59.

Cook, Edward (1985) *A Time Series Analysis Approach to Tree-Ring Standardisation*. PhD dissertation, University of Arizona.

Cook, Edward and Briffa, Kevin (1990) A comparison of some tree-ring standardisation methods. In E. R. Cook and L. A. Kairiukstis (eds) *Methods of Dendrochronology: Applications in the Systems Analysis*. Dordrecht: Kluwer Academic Publishers, pp. 153–161.

Cook, Edward and Pederson, Neil (2011) Uncertainty, emergence and statistics. In Malcolm Hughes, Thomas W. Swetnam and Henry F. Diaz (eds) *Dendroclimatology: Progress and Prospects*. Heidelberg; London; New York: Springer-Verlag, pp. 77– 112.

Cook, John, Nuccitelli, Dana, Green, Sarah, Richardson, Mark, Winkler, Bärber, Painting, Rob, et al. (2013) Quantifying the consensus on anthropogenic global warming in the scientific literature. *Environmental Research Letters*, 8 (2): 024024.

Coopmans, Catelijne and Button, Graham (2014) Eyeballing expertise. *Social Studies of Science*, 44 (5): 758–785.

Coopmans, Catelijne, Vertesi, Janet, Lynch, Michael and Woolgar, Steve (eds) (2014) *Representation in Scientific Practice Revisited.* Cambridge, MA: MIT Press.

Copenheaver, Carolyn, Goldbeck, Kyrille and Cherubini, Paolo (2010) Lack of gender bias in citation rates of publications by dendrochronologists: what is unique about this discipline? *Tree-Ring Research*, 66 (2): 127–133.

D'Arrigo, Rosanne, Wilson, Rob and Anchukaitis, Kevin J. (2013) Volcanic cooling signal in tree ring temperature records for the past millennium. *Journal of Geophysical Research: Atmospheres*, 118 (16): 9000–9010.

D'Arrigo, Rosanne, Wilson, Rob and Jacoby, Gordon (2006) On the long-term context for late twentieth century warming. *Journal of Geophysical Research: Atmospheres*, 111 (D3): 1–12.

D'Arrigo, Rosanne, Wilson, Rob, Lieperta, Beate and Cherubini, Paolo (2008) On the 'divergence problem' in northern forests: a review of the tree-ring evidence and possible causes. *Global Planetary Change*, 60 (3): 289–305.

Daston, Lorraine (1995) The moral economy of science. *Osiris*, 10: 2–24.

Daston, Lorraine and Galison, Peter (2007) *Objectivity*. New York: Zone Books.

Dear, Peter (2001) Science studies as epistemography. In Jay Labinger and Harry Collins (eds) *The One Culture: A Conversation About Science*, Chicago; London: University of Chicago Press, pp. 128–141.

Demeritt, David (2001) The construction of global warming and the politics of science. *Annals of the Association of American Geographers*, 91 (2): 307–337.

Dennis, Michael A. (1987) Accounting for research: new histories of corporate laboratories and the social history of American science. *Social Studies of Science*, 17 (3): 479–518.

Douglas, Mary ([1966] 2001) *Purity and Danger: An Analysis of Concepts of Pollution and Taboo*. New York: Frederick A. Praeger.

Douglass, Alexander (1937) Tree-ring work. *Tree-Ring Bulletin*, 4 (1): 3–6.

Duneier, Mitchell (2011) How not to lie with ethnography. *Sociological Methodology*, 41 (1): 1–11.

Edwards, Tamsin (2013) Twitter post, 21 October 2013, 6:16, https://twitter.com/flimsin/status/392278255476027393, accessed 8 September 2018.

Edwards, Paul (2010) *A Vast Machine: Computer Models, Climate Data, and the Politics of Global Warming*. Cambridge, MA: MIT Press.

Edwards, Paul, Mayernik, Matthew S., Batcheller, Archer L., Bowker, Geoffrey C. and Borgman, Christine L. (2011) Science friction: data, metadata, and collaboration. *Social Studies of Science*, 41 (5): 667–690.

Endfield, Georgina H., Veale, Lucy and Hall, Alexander (2015) Gordon Valentine Manley and his contribution to the study of climate change: a review of his life and work. *Wiley Interdisciplinary Reviews: Climate Change*, 6 (3): 287–299.

Esper, J., Frank, D. C., Büntgen, U., Verstege, A., Luterbacher, J. and Xoplaki, E. (2007) Long-term drought severity variations in Morocco. *Geophysical Research Letters*, 34 (7): L17702.

Esper, Jan, Büntgen, Ulf, Luterbacher, Jürg and Krusič, Paul J. (2013a) Testing the hypothesis of post-volcanic missing rings in temperature sensitive dendrochronological data. *Dendrochronologia*, 31(3): 216–222.

Esper, Jan, Schneider, Lea, Krusič, Paul, Luterbacher, Jürg, Büntgen, Ulf, Timonen, Mauri, et al. (2013b) European summer temperature response to annually dated volcanic eruptions over the past nine centuries. *Bulletin of Volcanology*, 75 (7): 1–14.

Farrell, Michael P. (2003) *Collaborative Circles: Friendship Dynamics and Creative Work*. Chicago: University of Chicago Press.

Felt, Ulrike, Fouché, Rayvon, Miller, Clark A. and Smith-Doerr, Laurel (2017) *The Handbook of Science and Technology Studies* (4th ed.). Cambridge, MA: MIT Press.

Feynman, Richard (1998) *The Meaning of It All: Thoughts of a Citizen Scientist*. Reading, MA: Perseus Publishing.

Fleck, Ludgwig ([1920] 1981) *The Genesis and Development of a Scientific Fact*. Chicago; London: University of Chicago Press.

Fletcher, Isabel (2014) Defining an epidemic: the Body Mass Index in British and American obesity research 1960–2000. *Sociology of Health & Illness*, 36 (3): 338–353.

Foucault, Michel (1980) *Power/knowledge: selected interviews and other writings (1972–1977)*. Edited by Colin Gordon. Harlow: Longman.

Fritts, Harold (1976) *Tree Rings and Climate*. London: Academic Press.

Frohlich, Xaq (2017) The informational turn in food politics: the US FDA's nutrition label as information infrastructure. *Social Studies of Science*, 42 (7): 145–171.

Garfinkel, Harold (1967) *Studies in Ethnomethodology*. New Jersey: Prentice-Hall Inc.

Geertz, Clifford (1973) *The Interpretation of Cultures: Selected Essays*. New York: Basic.

Gieryn, Thomas (1999) *Cultural Boundaries of Science: Credibility on the Line*. Chicago; London: University of Chicago Press.

Goffman, Erving (1967) *Interaction Ritual: Essays in Face to Face Behavior*. New Brunswick: Transaction Publishers.

Golinski, Jan (1998) *Making Natural Knowledge: Constructivism and the History of Science*. Cambridge: Cambridge University Press.

Golinski, Jan (1988) The secret life of an alchemist. In J. Fauvel, R. Flood, M. Shortland and R. Wilson (eds) *Let Newton Be!* Oxford: Oxford University Press, pp. 147–167.

González-Santos, Sandra and Diamond, Rebecca (2015) Medical and scientific conferences as sites of sociological interest: a review of the field. *Sociology Compass*, 9 (3): 235–245.

Goodwin, Charles (1994) Professional vision. *American Anthropologist*, 96 (3): 606–633.

Google Scholar (2019) Label: dendroclimatology, https://scholar.google.com/citations?hl=en&view_op=search_authors&mauthors=label%3Adendroclimatology&btnG=, accessed 24 May 2019.

Gould, J. S. (1965) Is uniformitarianism necessary? *American Journal of Science*, 263 (3): 223–228.

Gregory, Jane and Miller, Steve (1998) *Science in Public: Communication, Culture, and Credibility*. New York; London: Plenum Trade.

Gross, Neil and Fleming, Crystal (2011) Academic conferences and the making of philosophical knowledge. In Michele Lamont, Charles Camic and Neil Gross (eds) *Social Knowledge in the Making*. Chicago: University of Chicago Press, pp. 151–180.

Hacking, Ian (1983) *Representing and Intervening: Introductory Topics in the Philosophy of Natural Science*. Cambridge: Cambridge University Press.

Hagstrom, Warren (1982) 'Gift giving' as an organizing principle in science. In Barry Barnes and David Edge (eds) *Science in Context: Readings in the Sociology of Science*. Milton Keynes: Open University Press, pp. 21–34.

Haraway, Donna (1988) Situated knowledges: the science question in feminism and the privilege of partial perspectives. *Feminist Studies*, 14: 575–599.

Hatherly, David, Leung, David and MacKenzie, Donald (2008) The finitist accountant: classifications, rules and the construction of profits. In Trevor Pinch and Richard Swedberg (eds) *Living in a Material World: Economic Sociology Meets Science and Technology Studies*. Cambridge, MA: MIT Press, pp. 131–160.

Hess, David (2001) Ethnography and the development of science and technology studies. In Paul Atkinson, Amanda Coffey, Sara Delamont, John Lofland and Lyn Lofland (eds) *The Sage Handbook of Ethnography*. Thousand Oaks: Sage Publications, pp. 234–245.

Hevly, Bruce (1996) The heroic science of glacier motion. *Osiris*, 11 (1): 66–86.

Heymann, Matthias (2010) The evolution of climate ideas and knowledge. *Wiley Interdisciplinary Reviews: Climate Change*, 1 (4): 581–597.

Heymann, Matthias, Gramelsberger, Gabriele and Mahony, Martin (2017) *Cultures of Prediction in Atmospheric and Climate Science: Epistemic and Cultural Shifts in Computer-Based Modelling and Simulation*. Abingdon: Routledge.

Hirschman, Daniel and Reed, Isaac Ariail (2014) Formation stories and causality in sociology. *Sociological Theory*, 32 (4): 259–282.

House of Commons Science and Technology Committee (2010) *The Disclosure of Climate Data from the Climatic Research Unit at the University of East Anglia* (No. HC 387I), https://publications.parliament.uk/pa/cm200910/cmselect/cmsctech/387/387i.pdf, accessed 8 September 2018.

Hovland, Carl, Janis, Irving and Kelley, Harold (1953) *Communication and Persuasion*. New Haven: Yale University Press.

Howarth, Candice and Sharman, Amelia (2015) Labelling opinions in the climate debate: a critical review. *Wiley Interdisciplinary Reviews: Climate Change*, 6 (2): 239–254.

Howe, Joshua P. (2014) *Behind the Curve: Science and the Politics of Global Warming*. Seattle; London: University of Washington Press.

Hughes, Malcolm (2012) Comment to Mann et al. (2012) at Nature Geoscience. *ITRDBFOR* [international dendrochronology discussion forum]. 26 November 2012 16:42. [Not available online].

Hughes, Malcolm (2011) Dendroclimatology in high-resolution paleoclimatology. In Malcolm Hughes, Thomas W. Swetnam and Henry F. Diaz (eds) *Dendroclimatology: Progress and Prospects*. Heidelberg; London; New York: Springer-Verlag, pp. 17–36.

Hughes, Malcolm, Schweingruber, Fritz, Cartwright, D. and Kelly, P. M. (1984) July–August temperature at Edinburgh between 1721 and 1975 from tree-ring density and width data. *Nature*, 308: 341–343.

Hulme, Mike and Ravetz, Jerome (2009) Show your working: what 'ClimateGate' means. *BBC News*, 1 December 2009, http://news.bbc.co.uk/2/hi/8388485.stm, accessed 8 September 2018.

IPCC (2007) Summary for policymakers. In S. Solomon, D. Qin, M. Manning, Z. Chen, M. Marquis, K. B. Averyt, et al. (eds) *Climate Change 2007: The Physical Science Basis. Contribution of Working Group I to the Fourth Assessment Report of the Intergovernmental Panel on Climate Change*. Cambridge; New York: Cambridge University Press, www.ipcc.ch/site/assets/uploads/2018/02/ar4-wg1-spm-1.pdf, accessed 14 May 2019.

Jasanoff, Sheila (2010) Testing time for climate science. *Science*, 328 (5979): 695–696, http://science.sciencemag.org/content/328/5979/695, accessed 8 September 2018.

Jasanoff, Sheila (2006) Transparency in public science: purposes, reasons, limits. *Law and Contemporary Problems*, 69: 21–46.

Jeffrey, Bob and Troman, Geoff (2004) Time for ethnography. *British Educational Research Journal*, 30 (4): 535–548.

Jensen, Casper and Lauritsen, Peter (2005) Qualitative research as partial connection: bypassing the power–knowledge nexus. *Qualitative Research*, 5 (1): 59–77.

Jones, Phil (2016) The reliability of global and hemispheric surface temperature records. *Advances in Atmospheric Sciences*, 33 (3): 269–282.

Jones, P. D. and Lister, D. (2004) The development of monthly temperature series for Scotland and Northern Ireland. *International Journal of Climatology*, 24 (5): 569–590.

Josh 241 (2013) More battling. 'Bishop Hill' blog, http://bishophill.squarespace.com/blog/2013/10/21/more-battling-josh-241.html, accessed 27 May 2019.

Kamwendo, Thoko (2017) *Heuristics and Biases to Behavioural Economics: A Sociology of a Psychology for Euuor*. PhD thesis, The University of Edinburgh. www.era.lib.ed.ac.uk/handle/1842/25831.

Katz, Jack (2002) From how to why (part II). *Ethnography*, 3 (1): 63–90.

Katz, Jack (2001) From how to why (part I). *Ethnography*, 2 (4): 443–473.

Keenan, Douglas and Wilson, Rob (2015). Email correspondence, 13 March.

Knorr Cetina, Karin (2008) Objectual practice. In Massimo Mazzotti (ed.) *Knowledge as Social Order: Rethinking the Sociology of Barry Barnes*. Aldershot: Agate, pp. 83–98.

Knorr Cetina, Karin (1999) *Epistemic Cultures: How the Sciences Make Knowledge*. Cambridge, MA: Harvard University Press.

Knorr Cetina, Karin (1982) Scientific communities or transepistemic arenas of research? A critique of quasi- economic models of science. *Social Studies of Science*, 12 (1): 101–130.

Knorr Cetina, Karin (1981) *The Manufacture of Knowledge: An Essay on the Constructivist and Contextual Nature of Science*. Oxford; New York: Pergamon Press.

Kohler, Robert (2006) *All Creatures: Naturalists, Collectors, and Biodiversity, 1850–1950*. Princeton: Princeton University Press.

Kohler, Robert (2002) *Landscapes and Labscapes: Exploring the Lab–Field Border in Biology*. Chicago: University of Chicago Press.

Kohler, Robert E. (1991) Drosophila and evolutionary genetics: the moral economy of scientific practice. *History of Science*, 29 (4): 335–375.

Kohler, Robert and Vetter, Jeremy (2016) The Field in *A Companion to the History of Science*. Edited by Bernard Lightman. John Wiley & Sons, Ltd. Published.

Kuhn, T. S. (1963) The function of dogma in scientific research. In A. C. Crombie (ed.) *Scientific Change*. New York; London: Basic Books and Heineman, pp. 347–369.

Kuhn, Thomas ([1962] 1970) *The Structure of Scientific Revolutions*. Chicago: University of Chicago Press.
Kuklick, Henrika and Kohler, Robert (1996) *Science in the Field*. Chicago: University of Chicago Press. Originally published as a special issue in *Osiris* (1996) 11.
Kusch, Martin (2002) *Knowledge by Agreement: The Programme of Communitarian Epistemology*. Oxford; New York: Clarendon Press; Oxford University Press.
Lahsen, Myanna (2013a) Anatomy of dissent: a cultural analysis of climate skepticism. *American Behavioral Scientist*,57 (6): 732–753.
Lahsen, Myanna (2013b) Climategate: the role of the social sciences. *Climatic Change*, 119 (3–4): 547–558.
Lamont, Michele, Camic, Charles and Gross, Neil (2011) *Social Knowledge in the Making*. Chicago: University of Chicago Press.
Lanner, Ronald (2012) Comment to Mann et al. (2012) at Nature Geoscience. *ITRDBFOR*. 26 November 2012 03:48.
Latimer, Joanna and López, Daniel (2019) Intimate entanglements: affects, more-than-human intimacies and the politics of relations in science and technology. *Sociological Review Monograph Series*, 67 (2): 247–263.
Latour, Bruno (2005) *Reassembling the Social: An Introduction to Actor-Network-Theory*. Oxford; New York: Oxford University Press.
Latour, Bruno (2004) Why has critique run out of steam? From matters of fact to matters of concern. *Critical Inquiry*, 30 (2): 225–248.
Latour, Bruno (1999) *Pandora's Hope: Essays on the Reality of Science Studies*. Cambridge, MA; London: Harvard University Press.
Latour, Bruno (1987) *Science in Action: How to Follow Scientists and Engineers Through Society*. Cambridge, MA: Harvard University Press.
Latour, Bruno and Woolgar, Steven ([1979] 1986) *Laboratory Life: The Social Construction of Scientific Facts*. Princeton; Chichester: Princeton University Press.
Laudel, Grit and Gläser, Jochen (2007) Interviewing scientists. *Science, Technology and Innovation Studies*, 3 (2): 91–111.
Law, John (2008) Actor-Network Theory and material semiotics. In B. S. Turner (ed.) *The New Blackwell Companion to Social Theory* (3rd ed.). Oxford: Blackwell, pp. 141–158.
Lawrence, Christopher and Shapin, Steven (1998) *Science Incarnate: Historical Embodiments of Natural Knowledge*. Chicago: Chicago University Press.
Lynch, Michael (2017) STS, symmetry and post-truth. *Social Studies of Science*, 47 (4): 593–599.
Lynch, Michael (2013) Ontography: investigating the production of things, deflating ontology. *Social Studies of Science*, 43 (3): 444–462.
Lynch, Michael (2008) *Truth Machine: The Contentious History of DNA Fingerprinting*. Chicago: Chicago University Press.
Lynch, Michael (2000) Against reflexivity as an academic virtue and source of privileged knowledge. *Theory, Culture and Society*, 17 (3): 26–54.
Lynch, Michael (1985) Discipline and the material form of images: an analysis of scientific visibility. *Social Studies of Science*, 15 (1): 37–66.
MacKenzie, Donald (2006) *An Engine, Not a Camera: How Financial Models Shape Markets*. Cambridge; London: MIT Press.
MacKenzie, Donald, Muniesa, Fabian and Siu, Lucia (2007) *Do Economists Make Markets? On the Performativity of Economics*. Princeton: Princeton University Press.
MacKenzie, Donald and Spears, Taylor (2014a) 'The formula that killed Wall Street: the Gaussian copula and modelling practices in investment banking. *Social Studies of Science*, 44 (3): 393–417.

MacKenzie, Donald and Spears, Taylor (2014b) a device for being able to book p&l: the organizational embedding of the Gaussian copula. *Social Studies of Science*, 44: 418–440.

Mahony, Martin and Hulme, Mike (2018) Epistemic geographies of climate change: science, space and politics. *Progress in Human Geography*, 42 (3): 395–424.

Mann, Michael (2013a) Twitter post, 21 October 2013, 07:12, https://twitter.com/michaelemann/status/392292189176619008, accessed 8 September 2018.

Mann, Michael (2013b) Twitter post, 21 October 2013, 08:27, https://twitter.com/michaelemann/status/392311221371678720, accessed 8 September 2018.

Mann, Michael (2012) *The Hockey Stick and the Climate Wars: Dispatches from the Front Lines*. New York: Columbia University Press.

Mann, M. (2009) Science journals must be unpolluted by politics. *Wall Street Journal*, 31 December 2009, www.wsj.com/articles/SB10001424052748703478704574612400823765102, accessed 14 May 2019.

Mann, Michael, Bradley, Raymond and Hughes, Malcolm (1998) Global-scale temperature patterns and climate forcing over the past six centuries. *Nature*, 392 (6678): 779–787.

Mann, Michael, Fuentes, Jose D. and Rutherford, Scott (2012) Underestimation of volcanic cooling in tree-ring based reconstructions of hemispheric temperatures. *Nature Geoscience*, 5 (3): 202–205.

Mann, Michael E., Rutherford, Scott, Schurer, Andrew, Tett, Simon F. B. and Fuentes, Jose D. (2013) Discrepancies between the modeled and proxy-reconstructed response to volcanic forcing over the past millennium: implications and possible mechanisms. *Journal of Geophysical Research: Atmospheres*, 118 (14): 7617–7627.

Marcus, George (2009) Notes towards an ethnographic memoir of supervising graduate research through anthropology's decades of transformation. In James Faubion and George Marcus (eds) *Fieldwork Is Not What It Used to Be: Learning Anthropology's Method in a Time of Transition*. Ithaca: Cornell University Press, pp. 1–34.

Martin-Nielsen, Janet (2015) Ways of knowing climate: Hubert H. Lamb and climate research in the UK. *Wiley Interdisciplinary Reviews: Climate Change*, 6 (5): 465–477.

Mazanderani, Fadhila, Fletcher, Isabel and Schyfter, Pablo (2018) Introduction: talking STS. *Engaging Science, Technology and Society*, 4 (5): 179–182, https://estsjournal.org/index.php/ests/issue/view/10, accessed 8 September 2018.

Mazzotti, Massimo (ed.) (2008) *Knowledge as Social Order: Rethinking the Sociology of Barry Barnes*. Aldershot: Agate.

McIntyre, Steve (2016) Re-examining Cook's Mt Read (Tasmania) chronology. *Climate Audit*, 16 August, https://climateaudit.org/2016/08/16/re-examining-cooks-mt-read-tasmania-chronology/, accessed 8 September 2018.

McIntyre, Steve (2009) Gavin's guru and RCS standardization. *Climate Audit*, 4 October, https://climateaudit.org/2009/10/04/gavins-guru-and-rcs-standardization/, accessed 8 September 2018.

McKitrick, Ross (2013) Comment on 'Wilson on millennial temperature reconstructions'. 'Bishop Hill' blog, 21 October 2013, http://bishophill.squarespace.com/blog/2013/10/21/wilson-on-millennial-temperaturereconstructions.html, accessed 5 November 2018.

Melvin, Thomas, Grudd, Hakan and Briffa, Keith (2013) Potential bias in 'updating' tree-ring chronologies using regional curve standardisation: re-processing 1500 years of Torneträsk density and ring-width data. *The Holocene*, 23 (3): 364–373.

Merton, Robert (1978) The sociology of science: an episodic memoir. In R. K. Merton and J. Gaston (eds) *The Sociology of Science in Europe*. Carbondale: Southern Illinois University Press, pp. 3–14.

Merton, Robert (1972) Insiders and outsiders: a chapter in the sociology of knowledge. *American Journal of Sociology*, 78 (1): 9–47.

Merton, Robert (1968) *Social Theory and Social Structures*. New York: Macmillan.

Mialet, Hélène (2017) A singularity: where actor network theory breaks down an actor network becomes visible. *Subjectivity*, 10 (3): 313–328.

Mialet, Hélène (2012) *Hawking Incorporated: Stephen Hawking and the Anthropology of the Knowing Subject*. Chicago: University of Chicago Press.

Michael, Mike (2017) *Actor Network Theory: Trials, Trails and Translations*. London: Sage Publications.

Miller, Clark A. (2007) Democratization, international knowledge institutions, and global governance. *Governance*, 2 (2): 325–357.

Mol, Annemarie (2002) *The Body Multiple: Ontology in Medical Practice*. Durham; London: Duke University Press.

Monteiro, Eric, Almklov, Petter and Hepsø, Vidar (2012) Living in a sociomaterial world. In A. Bhattacherjee and B. Fitzgerald (eds) *Shaping the Future of ICT Research: Methods and Approaches*. IFIP Advances in Information and Communication Technology, vol. 389. Berlin; Heidelberg: Springer, pp. 91–197.

Montford, Andrew (2013) Wilson on millennial temperature reconstructions. 'Bishop Hill' blog, 21 October, http://bishophill.squarespace.com/blog/2013/10/21/wilson-on-millennial-temperaturereconstructions.html, accessed 8 September 2018.

Montford, Andrew (2012) Lonely old Mann. 'Bishop Hill' blog, 24 November, www.bishop-hill.net/blog/2012/11/26/lonely-old-mann.html, accessed 8 September 2018.

Montford, Andrew (2010) *The Hockey Stick Illusion: Climategate and the Corruption of Science*. London: Stacey International.

Mosher, Steve and Fuller, Thomas (2010) *Climategate: The CRUtape Letters*. Lexington: CreateSpace.

Muniesa, Fabian (2014) *The Provoked Economy: Economic Reality and the Performative Turn*. New York: Routledge.

Munro, Rolland (1997) Ideas of difference: stability, social spaces and labour of division. In Kevin Hetherington and Rolland Munro (eds) *Ideas of Difference: Social Spaces and the Labour of Division*. Blackwell Publishers/the Sociological Review, pp. 3–24.

Nehrbass-Ahles, Christoph, Babst, Flurin, Klesse, Stefan, Nötzli, Magdalena, Bouriaud, Olivier, Neukom, Raphael, et al. (2014) The influence of sampling design on tree-ring-based quantification of forest growth. *Global Change Biology*, 20 (9): 2867–2885.

Nerlich, B. (2010) 'Climategate': paradoxical metaphors and political paralysis. *Environmental Values*, 19 (4): 419–442.

Neyland, Daniel (2008) *Organizational Ethnography*. London: Sage.

Niiler, Eric (2017) Scientists did actually do some science at the march for science. WIRED, 24 April, www.wired.com/2017/04/scientists-actually-science-march-science/, accessed 16 November 2018.

Numbers, Robert (2009) *Galileo Goes to Jail and Other Myths about Science and Religion*. Cambridge, MA; London: Harvard University Press.

O'Neill, Onora (2013) What we don't understand about trust. TEDxHousesofParliament, June, www.ted.com/talks/onora_o_neill_what_we_don_t_understand_about_trust?language=en, accessed 8 September 2018.

O'Neill, Onora (2002) *A Question of Trust: The BBC Reith Lectures*. Cambridge: Cambridge University Press.

Østerlie, Thomas, Almklov, Peter and Hepsø, Vidar (2012) Dual materiality and knowing in petroleum production. *Information and Organization*, 22 (2): 85–105.

Pearce, Fred (2010) *The Climate Files: The Battle for the Truth about Global Warming*. London: Guardian Books/Random House.

Pearce, Warren, Grundmann, Reiner, Hulme, Mike, Raman, Sujatha, Kershaw, Eleanor Hadley and Tsouvalis, Judith (2017) Beyond counting climate consensus. *Environmental Communication*, 11 (6): 723–730.

Phillips, Christopher (2016) The taste machine: sense, subjectivity and statistics in the California wine world. *Social Studies of Science*, 46 (3): 461–481.

Phillips, Christopher (2019) *Scouting and Scoring: How we know what we know about baseball*. New Jersey: Princeton University Press.

Pilcher, J. (1990) Sample preparation, crossdating and measurement. In Edward R. Cook and L. A. Kairiukstis (eds) *Methods of Dendrochronology: Applications in the Environmental Sciences*. Dordrecht: Kluwer Academic Publishers, pp. 40–50.

Pollock, Neil and Williams, Robin (2016) *How Industry Analysts Shape the Digital Future*. Oxford: Oxford University Press.

Power, Michael (2005) The theory of the audit explosion. In Ewan Ferlie Laurence, E. Lynn and Christopher Pollitt (eds) *The Oxford Handbook of Public Management*. Oxford: Oxford University Press, pp. 326–344.

Power, Richard (2019) *The Overstory: A novel*. New York: W. W. Norton & Company.

Rabinow, Paul ([1977] 2007). *Reflections on Fieldwork in Morocco*. Berkeley; London: University of California Press.

Rafanell, Irene (2009) Durkheim's social facts and the performative model: reconsidering the objective nature of social phenomena. In Geoff Cooper, Andrew King and Ruth Rettie (eds) *Sociological Objects: Reconfigurations of Social Theory*. New York: Routledge, pp. 59–77.

Ramírez-i-Ollé, Meritxell (2019a) Friendship as a scientific method. *Sociological Review Monograph*, 67 (2): 299–317.

Ramírez-i-Ollé, Meritxell (2019b) Trust, scepticism and social order: a contribution from the sociology of scientific knowledge. *Sociology Compass*, 13 (2), e12653.

Ramírez-i-Ollé, Meritxell (2018) 'Civil scepticism' and the social construction of knowledge: a case in dendroclimatology. *Social Studies of Science*, 48 (6): 821–845.

Ramírez-i-Ollé, Meritxell (2015a) Rhetorical strategies for scientific authority: a boundary-work analysis of 'Climategate'. *Science as Culture*, 24 (4): 384–411.

Ramírez-i-Ollé, Meritxell (2015b) The social life of climate science. *Method Quarterly*, Issue 2, www.methodquarterly.com/2015/02/the-social-life-of-climate-science/, accessed 13 May 2019.

Revkin, Andrew (2009) Climate panel defends its findings and members, 'Dot Earth' blog, *New York Times*, https://dotearth.blogs.nytimes.com/2009/12/06/climate-science-panel-defends-climate-findings/, accessed 9 April 2017.

Reynolds, Joanna (2017) Missing out: reflections on the positioning of ethnographic research within an evaluative framing. *Ethnography*, 18 (3): 345–365.

Ribeiro Duarte, Tiago (2017) Mutual linguistic socialisation in interdisciplinary collaboration. In Luis Reyes-Galindo and Tiago Ribeiro Duarte (eds) *Intercultural Communication and Science and Technology Studies*, Cham: Springer International Publishing, pp. 55–78.

Røstvik, C. M. and Fyfe, A. (2018) Ladies, gentlemen, and scientific publication at the Royal Society, 1945–1990. *Open Library of Humanities*, 4 (1): 1–40.

Rutherford, Scott and Mann, Michael E. (2014) Missing tree rings and the AD 774–775 radiocarbon event. *Nature Climate Change*, 4 (8): 648–649.

Rydval, Miloš, Druckenbrod, Daniel, Anchukaitis, Kevin J. and Wilson, Rob (2016) Detection and removal of disturbance trends in tree-ring series for dendroclimatology. *Canadian Journal of Forest Research*, 46 (3): 387–401.

Rydval, Miloš, Loader, Neil J., Gunnarson, Björn E.,Druckenbrod, Daniel L., Linderholm, Hans W., Moreton, Steven G., et al. (2017) Reconstructing 800 years of summer temperatures in Scotland from tree rings. *Climate Dynamics*, 49 (9): 2951–2974.

Rydval, M., Druckenbrod, D. L., Svoboda, M., Trotsiuk, V., Janda, P., Mikoláš, M., et al. (2018) Influence of sampling and disturbance history on climatic sensitivity of temperature-limited conifers. *The Holocene*, 28 (10): 1574–1587.

Sarewitz, Daniel (2016) Saving science. *The New Atlantis*, Spring/Summer, www.thenewatlantis.com/publications/saving-science, accessed 8 September 2018.

Sarewitz, Daniel (2004) How science makes environmental controversies worse. *Environmental Science and Policy*, 7 (5): 385–403.

Schinkel, Willem (2016) Making climates comparable: comparison in paleoclimatology. *Social Studies of Science*, 46 (3): 374–395.

Schmidt, Gavin (2013) Twitter post, 21 October 2013, 13:39, https://twitter.com/climateofgavin/status/392384557523025920, accessed 8 September 2018.

Schneider, Lea, Smerdon, Jason E., Büntgen, Ulf, Wilson, Rob J. S., Myglan, Vladimir S., Kirdyanov, Alexander V. and Esper, Jan (2015) Revising midlatitude summer temperatures back to A.D. 600 based on a wood density network. *Geophysical Research Letters*, 42 (11): 4556–4562.

Schudson, Michael (2015) *The Rise of the Right to Know: Politics and the Culture of Transparency, 1945–1975*. Cambridge, MA: Harvard University Press.

Schweingruber, Fritz (1988) *Tree Rings-Basics and Applications of Dendrochronology*. Dordrecht; Boston: D. Reidel Publishing Company.

Schyfter, Pablo (2009) The bootstrapped artefact: a collectivist account of technological ontology, functions and normativity. *Studies in History and Philosophy of Science*, 40 (1): 102–111.

Scott, Pam, Richards, Evelyn and Martin, Brian (1990) Captives of controversy: the myth of the neutral social researcher in contemporary scientific controversies. *Science, Technology, and Human Values*, 15 (4): 474–494.

Scottish Pine Project (2019) Motivation, www.st-andrews.ac.uk/~rjsw/ScottishPine/motivation.html, accessed 15 May 2019.

Seibt, Johanna (2018) Process philosophy. *Stanford Encyclopedia of Philosophy* (Spring Edition), Edward N. Zalta (ed.), https://plato.stanford.edu/archives/spr2018/entries/process-philosophy/, accessed 8 September 2018.

Shackley, Simon and Wynne, Brian (1996) Representing uncertainty in global climate change science and policy: boundary-ordering devices and authority. *Science, Technology, and Human Values*, 21 (3): 275–302.

Shapin, Steven (2016) A taste of science: making the subjective objective in the California wine world. *Social Studies of Science*, 46 (3): 436–460.

Shapin, Steven (2010) *Never Pure: Historical Studies of Science as if It Was Produced by People with Bodies, Situated in Time, Space, Culture, and Society, and Struggling for Credibility and Authority*. Baltimore: Johns Hopkins University Press.

Shapin, Steven (2008) *Scientific Life: A Moral History of a Late Modern Vocation*. Chicago: Chicago University Press.

Shapin, Steven (2001) How to be antiscientific. In Jay A. Labinger and Harry Collins (eds) *The One Culture? A Conversation about Science*. Chicago: University of Chicago Press, pp. 99–115.

Shapin, Steven (1994) Trust, honesty and authority of science. In Ruth Ellen Bulger, Elizabeth Meyer Bobby and Harvey V. Fineberg (eds) *Society's Choices: Social and Ethical Decision Making in Biomedicine*. Washington, DC: National Academy Press, pp. 388–408.

Shapin, Steven (1992) Discipline and bounding: the history and sociology of science as seen through the externalism–internalism debate. *History of Science*, 30 (90): 333–369.

Shapin, Steven (1991) The mind is its own place: science and solitude in seventeenth-century England. *Science in Context*,4 (1): 191–218.

Shapin, Steven (1988) The house of experiment in seventeenth century England. *Isis*, 79 (3): 373–404.

Shapin, Steven (1984) Pump and circumstance: Robert Boyle's literary technology. *Social Studies of Science*, 14 (4): 481–520.

Shapin, Steven and Schaffer, Simon ([1985] 2011) *Leviathan and the Air-Pump: Hobbes, Boyle, and the Experimental Life*. Princeton: Princeton University Press.

Sharman, Amelia (2014) Mapping the climate sceptical blogosphere. *Global Environmental*, 26: 159–170.

Shea, James (1982) Twelve fallacies of uniformitarianism. *Geology*, 10 (9): 455–460.

Silver, Nate (2012) *The Signal and the Noise: Why So Many Predictions Fail – but Some Don't*. New York: Penguin Press.

Simmel, Georg (1906) The sociology of secrecy and of secret societies. *American Journal of Sociology*, 11 (4): 441–498.

Sismondo, Sergio (2009) *An Introduction to Science and Technology Studies*. Chichester: Wiley-Blackwell.

Skrydstrup, Martin (2017) Envisioning the future by predicting the past: proxies, praxis and prognosis in paleoclimatology. *Futures*, 92: 70–79.

Skrydstrup, Martin (2012) Modelling ice: a field diary of anticipation on the Greenland Ice Sheet. In Kirsten Hastrup and Martin Skrydstrup (eds) *The Social Life of Climate Change Models: Anticipating Nature*. London; New York: Routledge, pp. 163–182.

Smith, Steven (1997) The scientist and engineer's guide to digital signal processing, www.dspguide.com/whatdsp.htm, accessed 16 April 2019.

Spector, Malcolm and Kitsuse, John ([1977] 2001) *Constructing Social Problems*. New Brunswick: Transaction Publishers.

Speer, James H. (2010) *Fundamentals of Tree-Ring Research*. Tucson: University of Arizona Press.

Spinardi, Graham, Bisby, Luke and Torero, Jose (2017) A review of sociological issues in fire safety regulation, *Fire Technology*, 53 (3): 1011–1037.

St George, Scott (2013) Twitter post, 21 October 2013, 06:37am, https://twitter.com/scottstgeorge/status/392283501552484352, accessed 11 April 2019.

St George, Scott, Ault, Toby and Torbenson, Alex (2013) The rarity of absent growth rings in Northern Hemisphere forests outside the American Southwest. *Geophysical Research Letters*, 40 (14): 3727–3731.

Stoffel, Markus, Corona, Christophe, Guillet, Sébastien, Poulain, Virginie, Bekki, Slimane, Guiot, Joël, et al. (2015) Estimates of volcanic-induced cooling in the Northern Hemisphere over the past 1,500 years. *Nature Geoscience*, 8: 784–788.

Stokes, Marvin and Smiley, Tarah (1968) *An Introduction to Tree-Ring Dating*. Tucson: University of Arizona Press.

Sturdy, Steve (2007) Knowing cases: biomedicine in Edinburgh, 1887–1920. *Social Studies of Science*, 37 (5): 659–689.

The Guardian (2010a) Open letter: climate change and the integrity of science. *The Guardian*, May, www.theguardian.com/environment/2010/may/06/climate-science-open-letter, accessed 8 September 2018.

The Guardian (2010b) Climate science: truth and tribalism. *The Guardian*, www.theguardian.com/commentisfree/2010/feb/06/climate-science-truth-and-tribalism, accessed 8 September 2018.

Tsing, Anna (1993) *In the Realm of the Diamond Queen: Marginality in an Out-of-the-Way Place*. Princeton: Princeton University Press.

Vertesi, Janet (2012) Seeing like a Rover: visualization, embodiment, and interaction on the Mars Exploration Rover Mission. *Social Studies of Science*, 42 (3): 393–414.

Vol. 111.D 03103.

Von Storch, H. (2009) Good science, bad politics. *Wall Street Journal*, 22 December 2009, www.wsj.com/articles/SB10001424052748704238104574601443947078538, accessed 14 May 2019.

Wall Street Journal (2009) Editorial: rigging a climate 'consensus'. *Wall Street Journal*, 28 November 2009, www.wsj.com/articles/SB10001424052748703499404574559630382048494, accessed 13 May 2019.

Wallis, Roy (ed.) (1979) *On the Margins of Science: The Social Construction of Rejected Knowledge*. Sociological Review Monographs. Staffordshire: University of Keele.

Watts, Anthony (2012) Dendros stick it to the Mann. 'Watts Up With That?' blog 25 November 2012, http://wattsupwiththat.com/2012/11/25/dendros-stick-it-to-the-mann/, accessed 8 September 2018.

Weart, Spencer (2018) Climatology as a profession. *The Discovery of Global Warming. American Institute of Physics*. Updated February 2018, https://history.aip.org/climate/climogy.htm, accessed 8 September 2018.

Werret, Simon (2019) *Thrifty Science: Making the Most of Materials in the History of Experiment*. Chicago: University of Chicago Press.

Whitehead, Alfred (1929) *Process and Reality: An Essay in Cosmology*, New York: Macmillan. Critical edition by D. R. Griffin and D. W. Sherbourne, New York: Macmillan.

Wilson, Rob (2019) 'Home', The Scottish Pine Project website, May, www.st-andrews.ac.uk/~rjsw/ScottishPine/, accessed 23 May 2019.

Wilson, Rob (2013) Comment on 'Wilson on Millennial Temperature Reconstructions'. 'Bishop Hill' blog, 21 October, http://bishophill.squarespace.com/blog/2013/10/21/wilson-on-millennial-temperaturereconstructions.Html, accessed 15 August 2018.

Wilson, Rob (2012a) Comment on 'Lonely old Mann. 'Bishop Hill' blog, 26 November, www.bishop-hill.net/blog/2012/11/26/lonely-old-mann.html, accessed 8 September 2018.

Wilson, Rob (2012b) Comment on RealClimate, 'Global temperatures, volcanic eruptions, and trees that didn't bark', 8 February, www.realclimate.org/index.php/archives/2012/02/global-temperatures-volcanic-eruptions-andtrees-that-didnt-bark/, accessed 15 August 2018.

Wilson, Rob (2012c) Comment to Mann et al. (2012) at Nature Geoscience. ITRDBFOR. 25 November 2012, 20:43, [Not available online].

Wilson, Rob (2008) Dendrochronological investigations of Scots pine from the North-West Cairngorms Region, Scotland. Unpublished report prepared for the Rothiemurchus Estates, www.st-andrews.ac.uk/~rjsw/ScottishPine/PDFs/2008%20Pine%20Report.pdf, accessed 14 May 2019.

Wilson, Rob, Anchukaitis, Kevin J., Briffa, Keith R., Büntgen, Ulf, Cook, Edward R., D'Arrigo, Rosanne, et al. (2016) Last millennium Northern Hemisphere summer temperatures from tree rings: part II, spatially resolved reconstructions. *Quaternary Science Reviews*, 163: 1–22.

Wright Mills, Charles ([1959] 2000) *The Sociological Imagination*. Oxford: Oxford University Press.

Yale Forum on Climate Change & The Media (2010) A Yale Forum two-part special feature: scientists and journalists on 'lessons learned' (Pt. 1), www.yaleclimateconnections.org/2010/11/scientists-and-journalists-on-lessons-learned/, accessed 14 May 2019.

Zeitzoff, Thomas (2017) How social media is changing conflict. *Journal of Conflict Resolution*, 61 (9): 1970–1991.

Zuiderent-Jerak, Teun (2015) *Situated Intervention: Sociological Experiments in Health Care*. Cambridge, MA: MIT Press.

Index